AMAZING ANNOYING BIRDS

LIVING WITH AUSTRALIAN BRUSH-TURKEYS

ANN GÖTH

Natural

Publishing

NATURAL PUBLISHING

I am writing this book on the land of the Gadigal people of the Eora nation and have conducted my most recent BT research on the land of the Wallumattagal clan of the Dharug Nation. In the past, I have also studied BTs on the land of the Jagera and Turrbal peoples, as well as the Nalbo and Dallambara tribes. In the spirit of reconciliation, I acknowledge the Traditional Custodians of these lands and their connections to land, sea, and community. I pay my respect to their Elders, past and present, and extend that respect to all Aboriginal and Torres Strait Islander peoples today.

ISBN 978 0 6486037 02

FOREWORD BY DARRYL JONES

Something astonishing is underway in the suburbs of many Australian towns and cities right now. Something disruptive, disturbing and sometimes disastrous. But also fascinating, remarkable and revolutionary. The relentless advance of the brush-turkey, determinedly reclaiming its place in the grand scheme of things and reminding humans who so nearly wiped them out entirely that times have changed. This time it is us on the defensive, desperately trying to find a way to live with these ecosystem engineers. As Ann Göth shows very clearly in this book, the arrival of these clever, resourceful, enterprising birds in your backyard can be ominous or fascinating. But one thing is certain; they cannot be ignored.

Ann is the ideal person to write the story of these amazing, annoying birds. A scientist with a long and accomplished history of detailed research on brush-turkeys and related megapodes, she has also investigated their arrival and exploits in urban areas. She has experienced first-hand their destruction and the anguish they have wrought. She has shared in the heartbreak and exasperation.

But she has also witnessed the wonder and awe they can generate when the hidden details of their almost unbelievable lives are explained. The remarkably sophisticated process of constructing an incubator the size of a small car out of dirt and leaves. Of females who produce lots of enormous eggs only to abandon them deep inside a compost heap. The Herculine task of the hatchlings which have to dig through a metre of soil to take their first breath of fresh air. The seemingly impossible odds they face during their first days fending for themselves in a dangerous world.

Every aspect of the strange and fractious relationship between people and brush-turkeys is covered in this delightful book. It is scientifically sound as well as understandable by anyone. Whether you have experienced the impact of these birds or are simply interested in finding out what on earth is going on, this book will not disappoint.

Prof. Darryl Jones, Brush-turkey expert and award-winning author of several books on birds

PRAISE FOR AMAZING ANNOYING BIRDS

This is a comprehensive, informative, instructive, attractive, and at times amusing, well-balanced introduction to, and account of, one of the world's most intriguing and potentially (to gardeners) annoying bird species. Anyone and everyone sharing their habitat with the Australian brush-turkey should have this delightfully illustrated volume to hand. An appreciation of the bird's remarkable life history should alleviate much frustration caused by it.

-- **Dr. Clifford B. Frith OAM**, author of 16 books on birds, including two each on bowerbirds and birds of paradise.

Ann's book is full of interesting and practical information. A great attribute is her humorous and chatty style which makes even the technical parts (which would be the 'boring bits' from other authors) so readable that one can't skip them! The illustrations are superb too.

This will be an excellent book to recommend for our rescuers, carers, and the many members of the public who ask for our help with all aspects of living with brush-turkeys. Thank you, Ann, you have made our job so much easier with this long-needed text!

-- **Bev Young OAM**, Senior Volunteer and Community Educator with Sydney Wildlife Rescue

What fascinating birds. I remember first seeing them years ago while camping, and a bit too friendly, even in that non -urban area. Ann's book is readable and fun and should appeal to anyone with an interest in birds. Maybe she can change an opinion or two?

-- **Judy Harrington,** BirdLife Australia

After spending half her life studying the enigmatic Australian brush-turkey, Dr Ann Göth has finally hatched the egg and put pen to paper detailing her and other people's experience with the iconic birds that have become increasingly common in the urban areas of eastern Australia from Sydney northwards. Written in layman's language understandable to all, years of research explain the intimate habits of brush-turkeys and why they have become so numerous in bushland and suburban gardens in and around our cities. Love them or hate them, this is a must-read for any people interested in urban wildlife or simply getting brush-turkeys out of their garden!

-- **David Booth,** Honorary Associate Professor in Zoology, University of Queensland.

I wished I lived in Australia instead of Europe: after reading Ann Göth's book about the fascinating behaviour of brush-turkeys and their chicks. I would definitely lure them into my garden and spend day after day watching them instead of getting rid of them.

-- **René Dekker,** Author of "The Megapodes", Leiden, Netherlands.

Dr Göth's new book is a veritable encyclopaedia of brush-turkey facts, and stories, all based on her extensive experience with these much beloved and hated birds. Interviews with professionals dealing with brush-turkeys and lovely hand-drawn illustrations are a bonus. It is a must-read for anyone interested in this truly unique and fascinating species, whether to learn more about them or to just figure out how to get them out of the garden.

-- **Matthew Hall,** Ecologist, PhD on brush-turkeys

Ann Göth has written a charming little book on the wonders of the brush-turkey of Australia. It delves into the interesting details of its unique mound-building behaviour, which should bring delight to every reader. Those who live with this bird and currently consider it to be a nuisance may be enlightened enough to welcome it in their gardens. If not, Ann provides advice on how to discourage them.

-- **Roger Seymour,** Emeritus Professor, The University of Adelaide

This book contains so much more than I could have imagined, and it really reminded me that we share our city with an amazing animal. It is an entertaining and informative journey into the extraordinary biology of one of Australia's most unique birds and our relationship with them. While their approach to life means that brush-turkeys are much maligned by some, the tales of their exploits confirm their place as a unique Australian icon: hardworking, unassuming, and self-sufficient. You will look at these birds differently after learning about what they do, why they do it, and how they do it. And why we need to learn to live with them.

-- **Professor Dieter Hochuli,** Professor, The University of Sydney

This book contains four sections

Section 1
The Annoying Conflict
Chapters 1-5

Section 2
The Amazing brush-turkey
Chapters 6-14

Section 3
The Power of Knowledge
Chapters 15-16

Section 4
Living Harmoniously
Chapters 17-18

Contents

Introduction: Urban conflicts with unusual birds

Nowhere else in the world other than Australia will you find an ancient bird that decides to build a huge, messy mound of dirt and leaf litter in a tidy suburban backyard. These birds divide the nation: frustrated gardeners and land managers full of consternation versus those people who love brush-turkeys and are full of admiration for a bird that has adapted so well when moving from the shadows of the rainforest into our suburbs.

Imagine you have invested lots of money and time into creating a lovely garden or vegetable patch when all of a sudden a big bird moves in and decides to rake all your plants, mulch, watering system and solar lights into one big mound. It may also add the garden gnome. This certainly is a reason for being upset, and not many gardeners share the brush-turkey's version of an ideal landscape.

Such reshaping of gardens in brush-turkey style has been increasing in suburban and urban dwellings along Australia's East Coast these past two decades, and it has started the outbreak of a suburban war: gardeners against brush-turkeys, as well as those people who love brush-turkeys against those who loathe them. Some see the rise in native brush-turkeys as a good-news story about wildlife returning to the city, and they welcome the opportunity to watch native wildlife at ground level and so close to home. Others are very unhappy about sharing their property with such large birds that wreak expensive havoc in their carefully manicured garden; they may even view the birds' mound-building and raking as 'malicious intent' to destroy their backyards. We are currently witnessing a highly unusual urban wildlife conflict that is evolving daily and, in Sydney, this conflict is even a hot media topic.

When I started researching brush-turkeys in 2002, I could not get a permit to collect the eggs for artificial incubation anywhere in Sydney because the NSW National Parks and Wildlife Service office declared these native birds were "too rare". Instead, I had to venture north to the Central Coast, where they were easier to find due to the density of bushland and proximity of national parks. Not long after that, however, the media started reporting about these birds wreaking havoc in the northern suburbs of Sydney and, every year that followed, the birds appeared to be moving further south, eventually crossing the wide Parramatta River that divides the city in half. They recently reached the famous Bondi Beach in Sydney's east and may soon conquer the inner-city Royal Botanic Garden, right on the doorstep of the Sydney Opera House and Sydney Harbour Bridge.

The other big Australian city where brush-turkeys trash suburban gardens is Brisbane. When Professor Darryl Jones, the 'brush-turkey whisperer', moved there in the early 1980s, there were very few brush-turkeys in Brisbane, and he had to venture south to Mt Tamborine to study them[1]. Over the next 20 years, Darryl witnessed their population increase within Brisbane by a staggering 700%[2]. While the brush-turkey war started years earlier in Brisbane than it did in Sydney, many people there are still not willing to accept garden havoc à la brush-turkey style.

Section 1 *The Annoying Conflict* starts by giving more insight into this new urban-wildlife conflict: people's most common complaints (Chapter 1) and how the conflict is portrayed in Australia's media (Chapter 2). In Chapter 3, I provide some insider views from people who deal with this conflict daily, including a local council officer, a wildlife rescue organisation volunteer and parklands staff. Whether brush-turkeys are invading, intruding on or returning to our suburbs will be explored in Chapters 4 and 5.

As with any conflict, solutions can only be found if we know why the conflict is occurring for both parties. After all, brush-turkeys have needs, too. Chapters 6-14 make up Section 2 *The Amazing Brush-turkey* and will explain more about what makes brush-turkeys tick: we will delve deeper into questions such as why they build mounds, how and when they do it, and what makes them so unique within our native fauna. You may be intrigued to learn about their unconventional mating and social behaviours as well as their highly unusual way of raising (or not raising!) their kids and the chicks' behaviour in the soil digging themselves out of the mounds. Across the book we'll explore a range of topics, from what brush-turkeys eat to how their eggs cope with being buried in the soil for seven weeks.

In Section 3 *The Power of Knowledge*, we examine how these birds have been a part of Aboriginal traditional beliefs for millennia and what those beliefs are teaching modern researchers (Chapter 15) before looking at studies by modern scientists (Chapter 16).

Section 4 *Living in Harmony* suggests methods you can apply if you are keen to deter brush-turkeys from your property (Chapter 17), followed by how you can help brush-turkeys (Chapter 18). The latter includes answers to your questions on what to do if you find an injured brush-turkey that seems to need help, and how you can support these birds in other ways.

Figure 1. A suburban brush-turkey on his incubation mound (drawing not to scale).

So far, I have used the common name *brush-turkey,* but many people know them as either *bush-turkey* or *scrub-turkey*. In the past, they were also called *tallegalla* or even *New Holland vulture*. Their Latin name is *Alectura lathami (see Chapter 16 for more detail)*. The old-fashioned word 'brush' describes any woody vegetation and, apparently, those choosing this name thought that brush-turkeys only occurred in this type of environment. Little did they know how adaptable brush-turkeys are and that they can also live in busy cities where proper brush is rare among the roads and manicured parks and gardens.

Whatever name you prefer for these birds is OK. **For ease of reading in this book, I will use the shortcut BT for brush-turkey, bush-turkey or scrub-turkey from now on. Let's start by exploring the common complaints people have about BTs.**

Section 1
The Annoying Conflict

CHAPTERS 1-5

Top: Brush-turkey mound in the city under washing line. Photo A. Göth

Left: Uninvited brush-turkey inspecting house. Photo: E. Mann.

1

COMMON COMPLAINTS

WE'RE TALKING HUNDREDS OF DOLLARS, MATE!

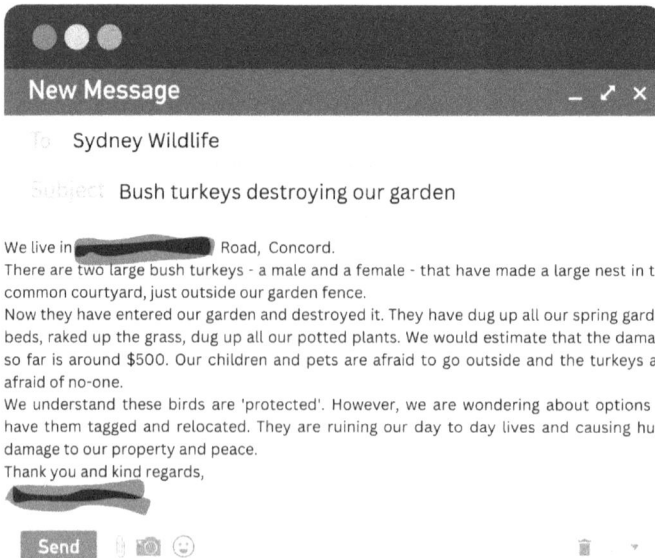

Figure 2. Email to Sydney Wildlife Rescue from a Sydney resident.

Residents along the Australian East Coast are often distressed about problems BTs cause when they either feed, roost or build mounds in people's backyards and nearby parks. In my regular public talks on BTs for local councils, I sometimes include questionnaires to my audience about their most pressing issues with these birds. The following chart shows the responses from 165 people who live in either Sydney, Parramatta or 750 kilometres (466 miles) further north in Byron Bay, New South Wales.

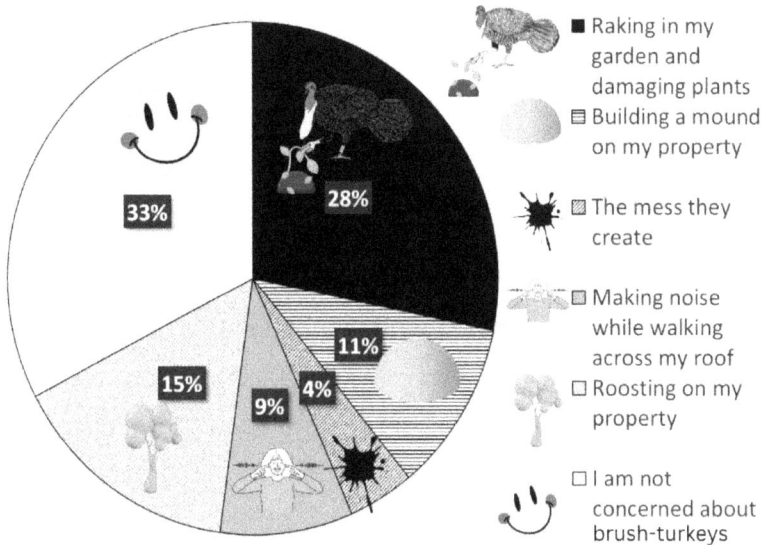

■ Raking in my garden and damaging plants

⊟ Building a mound on my property

▨ The mess they create

▨ Making noise while walking across my roof

☐ Roosting on my property

☐ I am not concerned about brush-turkeys

Figure 3. A pie chart shows people's responses (as %) to a survey about issues with BTs. Each of the 165 participants could choose multiple answers.

While 33% of people had no concerns about BTs, the remaining 67% did. The destruction of garden plants, including flowers and vegetables, was the most common complaint, followed by the birds building a mound. People were also worried about the mess these birds make raking the ground and the noise they cause when walking along tin roofs early in the morning. BTs roosting together in trees at night was an additional concern as this produces mess underneath the tree.

Let's look at each of these complaints in detail, plus add some additional ones I have come across more recently. These include the injuring of domestic chickens, attacking dogs and other odd complaints.

Destroying plants and gardens

BTs spend hours raking the ground with their strong legs and large claws to search for food, such as invertebrates, tubers and fruit. Our gardens' mulched and watered soil is rich in worms and other critters, all ideal BT food. While feeding, these birds inadvertently damage smaller plants and vegetables or expose the roots of larger plants, and may pull out watering systems, solar lights and other garden ornaments. They may occasionally nip on leafy greens or steal some fruit, but the main damage is done through their raking. The damage from raking can amount to hundreds of dollars, and it can be incredibly frustrating when several of these birds choose your backyard to feed.

People who love to grow greenery in pots sometimes complain about BTs uprooting those plants, occasionally upending or damaging the pots. These birds like the underground tubers of plants, which are the thickened parts of the roots, and some become specialists in unearthing those from pots.

Building mounds

The construction of a car-sized incubator made of leaf litter, mulch and other plant material brings considerable changes to the landscape. By raking together all this material, the male BT can cause the same damage to ornamental and vegetable gardens as described above. In addition, he strips the area of topsoil, damages watering systems and creates a slip hazard during the first few weeks of building when he builds a trail of leaf litter leading to the mound (see Chapter 8). The mound inevitably attracts several females, which means more BTs scratching for food in the garden, potentially increasing the damage. The mound also produces several chicks over the breeding season, further increasing the local BT population (though few survive — see Chapter 13).

Making noise while walking across roofs

BTs start their day as soon as the sun rises and they are creatures of habit. Every morning, they descend from their roost tree and travel the same or similar route to where they feed. If this route takes them across corrugated-iron roofs, you will hear the noise of big feet walking above you every morning, and not everyone appreciates such a wake-up alarm.

Roosting

At night, BTs go up high into a tree to roost and like to do so in a group. In the city, such groups can be large, up to 20 birds or so, because these birds have found food directly or indirectly provided by humans (see Chapter 5). If the BTs sit in a tree above your house or driveway at night, the results of their digestion end up underneath, and some people must clean up BT poop from their driveway every day to avoid slip hazards or an unsightly mess.

Attacking and/or killing my chickens

There are many areas where domestic chickens and BTs live together harmoniously but, unfortunately, BT assaults on chickens are happening, too. They are frequently observed in suburban areas where people love keeping chickens. Initially, the BTs are attracted to the food provided for the chickens but, at times, some males attempt to mate with the hens. The latter get injured and traumatised and occasionally die from their injuries. There is no chance the male BT will successfully sire any offspring, as the two species are too different, but this interaction can be distressing for the chickens and their owners.

A blog from the Northern Beaches area in Sydney reveals that only certain aggressive males attack and sometimes kill chickens. This blogger noted that for many years the BTs were hanging around the backyard with the chickens, feeding on their food and occasionally stealing an egg to eat. But they never attacked the chickens and were chased away by one of the boss chooks. This all changed when a male BT in full breeding mode moved in during the summer breeding season. He tried mating with the chooks and, while doing so, inflicted severe injuries that killed some of them. Attacks on chickens are predominantly undertaken by testosterone-filled BT males during the breeding season. The blog writer had to fence in his chickens, at least while the aggressive male was present[1].

<u>Attacking my dog</u>

A male BT who owns a mound will vigorously defend the eggs bestowed upon him by trying to chase away any potential predators, especially those with a tail. In the wild, these are often goannas (large lizards) and snakes. You can watch a video on YouTube showing a goanna burying its head into a mound to steal an egg, with its tail emerging at the top of the mound. The BT mound owner repeatedly pecks the goanna's tail and vigorously kicks dirt at the reptile[2]. Males use the same method to successfully chase away snakes and even harmless blue-tongued lizards (see Chapter 11).

Hence, the long, thin tail is the part of a predator that a male BT will first attack. Another viral YouTube video shows a dog in a backyard with its leash on the ground still attached to its collar. The male BT runs towards the leash and attacks it by pecking because, in his mind, it resembles a tail. The BT doesn't try to take the dog for a walk, as the videographer wrongly assumes; it just mistakes the leash for a predator's tail[3].

When you take your dog for a walk near a male with a mound, the BT will occasionally chase the dog, especially the tail. Smaller dogs and those with long, thin tails are more likely to become the subject of BT pursuits than larger ones or those with severed tails.

<u>Additional complaints</u>

One concern raised occasionally is that BTs are ugly to look at: people prefer cute and cuddly animals. Let's consider why they look the way they do. Their bald red head is practical when they must stick them deep down into the soil to test the mound's temperature (see Chapter 8); a feathered head would get muddy and make little sense for that activity. Their long legs and strong claws are essential for all that raking and digging, and the males' bare yellow wattle attracts females and/or discourages other males. Black plumage provides perfect camouflage in the dark undergrowth of the rainforest, their natural habitat. So, from a BT's perspective, their looks are a reflection of their lifestyle.

Certain people, especially small children, find BTs intimidating and sometimes fear them. It doesn't help that some urban BTs have become very accustomed to people and either approach them or at least don't run away quickly when approached. I can assure readers that there is no report of a BT ever attacking a person of any size; they are just a big and bold bird stalking around.

A last additional complaint is that BTs raid bags, bins and compost heaps to find food. They certainly are adept at finding convenient food sources, especially as they eat almost everything. They have not become quite as bin-dependent as the white ibis, also lovingly known as the 'bin-chicken', but BTs do tend to scavenge through plastic bags, bins or compost heaps that are not covered or zipped up. If you leave your open bags unattended at popular picnic spots, there is a high chance it may get raided by a BT!

With all these complaints from the public in mind, how are BTs portrayed in the Australian media? Do most journalists portray these birds as unpopular?

INTERVIEW

MATT HALL, THE UNIVERSITY OF SYDNEY

WHAT IS YOUR ROLE?

66 I am en ecologist. In 2022, I completed my PhD on urban brush-turkeys at The University of Sydney. 99

WHAT ARE THE COMMON COMPLAINTS PEOPLE HAVE ABOUT BRUSH-TURKEYS IN THE CITY?

66The most common complaints I receive about brush-turkeys revolve around their digging and raking habits. Brush-turkeys will rake through soil and leaf litter while foraging. This behaviour can easily ruin carefully cultivated garden beds, throwing mulch and soil about carelessly. Male brush-turkeys go even further during the breeding season and can shift tonnes of soil and leaf litter to build their mounds, often tearing up plants. Many people see the mounds themselves as an eyesore, particularly when they appear unexpectedly on top of a nicely manicured lawn.

Complaints about brush-turkeys aren't just limited to their raking, however. They have also been known to chase pets, steal pet food, dirty pool water, and attack backyard chickens. The patter of brush-turkeys walking across the roof is an unwelcome alarm clock for many a suburban resident, and there's also the unpleasant smell of droppings if people are unlucky enough to live near a roost.

Irate gardeners will use various techniques to try and keep industrious brush-turkeys out, from chicken wire and fencing to DIY scarecrows and strategically placed mirrors. Often these techniques meet with mixed success as brush-turkeys stubbornly return each morning to resume their raking despite residents' best efforts. Importantly, these experiences are not universal. Not all brush-turkeys are a nuisance. Many suburbanites are delighted to see a new bird species. Others report much more positive or benign interactions, showing that, in some cases at least, living alongside brush-turkeys is possible.99

2

BRUSH-TURKEYS IN THE MEDIA

TRIFFID-LIKE CREATURES? WORSE THAN CANE TOADS?

If you live on Australia's East Coast, there is no shortage of media articles about BTs to amuse, annoy or entertain. An "avian apparition" showing up on a journalist's balcony next to one of the busiest streets in Sydney sparked the idea for a five-page article in the esteemed *Australian Geographic* magazine[1]. The article explains why a "prehistoric-looking creature" has conquered the "concrete jungle of Neutral Bay rather than the deep shadows of the forest" and suggests that these birds are recolonising their former haunts rather than invading.

When writing about BTs, many newspapers use headlines with negative and incorrect implications, such as that BTs are "invading" Sydney, but these articles at least give voice to both the BT haters and lovers[2]. Headlines are there to attract readers, so it is no wonder that an otherwise informative article about a BT observed eating bandicoot roadkill was titled "Brush turkeys turn carnivorous in Sydney suburbs"[3]. A *New Daily* article went even further to lure in readers. Not only did it employ a sensationalist headline ("The turkey army that's invading suburban Sydney"), it used phrases in the first few paragraphs to clearly attract those not so keen on BTs: "How far will these triffid-like creatures spread?" and "This sudden expansion of territory has prompted fears that – like the cane toad, ibis and Indian myna – they're relentlessly moving down the eastern seaboard".

However, the journalist deserves some credit for providing more substantial and positive information about BTs in the rest of the article, including confirming that they are "returning" and not "invading"[4].

But BTs don't often get a fair go in the media, as can be seen by these other headlines:

- "Are brush turkeys becoming a plague?"[5]

- "Man v bird: the brush turkey battle"[6]

- "Brush turkeys are back in force and driving North Sydney gardeners mad with their backyard antics"[7]

- "Back from the bush: turkeys hit Sydney backyards"[8]

- "Dealing with brush turkeys: The bird with the bad reputation"[9]

Figure 4. Collage of newspaper headlines about brush-turkeys

However, not all BT publicity has been entirely bad. A variety of articles have investigated why these birds are booming in urban areas[10], how they are crossing Sydney Harbour given they can't fly well[11], why they are likely to come to a Sydney suburb near you[12] and what they are up to in Summer Hill, an inner-west suburb of Sydney, after being sighted there recently[13].

If you look hard and wide, you may even come across an article from a courageous journalist who dares to start with an extremely positive headline, such as "Five reasons to love brush turkeys"[14]. However, this was the only positive example I could find. But the media does call out people who harm BTs on purpose, such as in a recent poisoning incident in Sydney ("Fowl Play: brush turkeys possibly 'poisoned' at Allan Border oval")[15] or when a suffering BT in Byron Bay was sighted with an arrow through its wing ("Locals distressed as someone is violently attacking the protected brush turkeys of Byron Bay")[16]. The poisoning incident prompted several media outlets to look at the special features of BTs, including in an ABC radio interview with Dr John Martin, who is involved in the Big City Bird App research project (see Chapter 18)[17].

Opinion pieces about BTs can be quite an amusing read. One journalist described how he had previously disliked a possum that had eaten his garden plants, but now redirected his extreme dislike to the BT who both ate AND dug up his plants. As a result, he started looking up BT recipes on the internet[18]. This notion was quickly refuted by the writer of another opinion piece who suggested that instead of cooking BTs, these birds should be "admired for adding to the rich tapestry of Sydney's wildlife"[19].

Journalists can be very creative when coming up with nicknames for BTs but, unfortunately, names with negative connotations dominate. For example, these birds have been called "bourgeois birds", "red-necked game birds", "scavenging nuisances" and "triffid-like creatures". The figure below contains all the expressions I could find for BTs in the media, and you may have additional ones. If you do, please send these to me so I can add them to the word cloud. Positive ones are especially welcome, as they are clearly missing in most articles I could find.

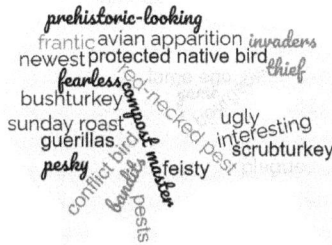

Figure 5. Words used to describe brush-turkeys in the media, extracted from various news articles.

The international media pays less attention to BTs' antics. Still, the UK's *Daily Mail* thought it was worth an article when an "angry" BT caused a "disaster" at a Sydney wedding: he gate-crashed the wedding's photoshoot while fighting with a kookaburra right in front of the bridal party. Now, that's a disaster indeed[20]!

Overall, we must remember that the media likes to be sensationalist and dwell on the irritating and annoying side of things, partly explaining the negative image BTs have in the news. There are clearly many people annoyed with BTs. What do the land and wildlife managers have to say about dealing with angry people and the BTs causing havoc in the city?

3

BRUSH-TURKEY MANAGERS

CASE STUDIES OF URBAN TURKEY CONFLICTS

L et's meet some of the dedicated people managing the urban wildlife con-flicts that BTs cause all along the East Coast of Australia. In my interviews with them, they spoke about the many issues they face. These managers are from local councils, the Sydney Olympic Park Authority, the NSW National Parks and Wildlife Service and Sydney Wildlife Rescue. I apologise to the managers further north, as Sydney is over-represented in this chapter simply because it is my hometown and I have more connections here. I have been told the situation is similar further north, even though people have had more time to get used to BTs recolonising those areas.

MEETING

TEAMS MEETING WITH: **Stephanie Martin, Ranger, Sydney North Area, Greater Sydney Branch, NSW National Parks and Wildlife Service**

SUBJECT: **Issues caused by brush-turkeys in her area and how NPWS deals with them.**

DATE: **27.10.2023**

NOTES:

She and her colleagues deal with a wide range of BT issues, from people loving them too much to those who want them gone. In several apartment blocks where BTs build a mound, some owners like to watch, feed, and protect them, and others don't like the mess and would like them removed. Stephanie and fellow rangers are often caught in the middle and seem to manage neighbourhood conflicts as much as the BTs. All they can do is advise on how to live with BTs by making modifications to their garden areas, including keeping food away, or how people can discourage the birds from mound building in the first place.

Some members of the public have suggested that the BTs be removed from the Biodiversity Conservation Act as a protected species, and that there should be a control program for these birds because of the burden they cause to some residents. Many residents become frustrated at the mess BTs make and are worried about the tripping hazards they cause. BTs commonly roamed residential areas prior to urbanisation, which some may not accept.

A lot of the rangers' work focuses on reactive issues. Many people think that if they call up the NPWS, a ranger will immediately catch the BT and take it away. It wasn't there before, after all. They find it hard to accept that this is not what happens, and the rangers explain to members of the public that there are only two options: adapt to living with BTs (by either making modifications to their garden, or deter the BTs from

MEETING

moving in) or, in expediential circumstances, where significant property damage occurs, euthanise them. People don't usually want them killed, just removed.

Euthanasia is not the desired outcome and is often a band-aid solution to managing the problem. It is used in very specific cases where human health and safety is at risk. The trouble is, once the individual bird is gone, more BTs will move in and take over the territory but they know that just translocating the birds to another area is a death sentence for them as well. Other BTs will already live in those areas and drive the intruder away, and often, they get killed on roads when trying to return home. Rangers much prefer educating people about how to deter the birds from building another mound next year and advising people how they can implement a landscape management plan. This advice includes trimming tree canopies to remove shade above the mound and removing mulch or food sources. The rangers need to point out that once they have given this advice, it is the landholders' job to adhere to it. Rangers can't come back repeatedly to deal with the same BT issue on the same property every year; they don't have the time.

Stephanie and her rangers are dealing with new issues every day. She hopes that as time passes, people will be more willing to adjust their lifestyle and accept that these birds are here to stay, just like it happened in Brisbane 20 years ago. Maybe in time, residents can learn to live with BTs and appreciate them for the fascinating, fastidious and devoted birds they are.

INTERVIEW

MICHELLE GREENFIELD
LANE COVE COUNCIL, SYDNEY

WHAT IS YOUR ROLE?

66 I am the Bushland Interpretation Officer for Lane Cove Council. 99

WHAT ISSUES DO YOU HAVE WITH BRUSH-TURKEYS IN YOUR COUNCIL AREA?

66 Australian brush-turkey numbers have exploded in our council area over the last ten years. We now find them in almost all of Lane Cove's bushland reserves or parks and many private gardens, where we deal with most of the conflicts these birds cause. People either love or hate them, and there seems to be no golden middle ground. Those who dislike them often do so because they rake up their gardens and build mounds in their backyards. We only recently had to respond to a distressed owner of a suburban house who could not keep the washing or backyard clean because up to 30 brush-turkeys chose to roost above her washing line. As you know, their droppings are of considerable size. We also had reports of mounds either blocking a footpath or being built near a path, with the leaf litter causing a tripping hazard. A few days ago, a lady called me worried the turkey would attack her small pet dog.

It was easy to respond to the lady with the dog as we explained that the males sometimes do that near the mound to protect their eggs from natural predators that have a tail, such as goannas. She has to carry her dog past the mound, and it will be okay beyond that. The other issues take more time to deal with. We often find that residents feed the birds, deliberately or otherwise, and we try to discourage that. Where mounds block paths, we check whether we can shift the mound across or create a detour and consider educational signs. The challenges we face with brush-turkeys are often new, but the most important point to bring across to our residents is that these birds are a protected native species. For the last seven years, we have also engaged Dr. Göth to present educational talks about brush-turkeys, which are always well received. 99

MEETING

MET WITH: Bev Young OAM, Senior Volunteer and Community Educator with Sydney Wildlife Rescue

SUBJECT : Brush-turkeys on the 24hr rescue hotline (02 9413 4300)

NOTES :

This hotline is for people to ring to get help for injured and orphaned wildlife found in the Sydney area. During the last few years, they have been getting many calls about brush-turkeys.

Quite a few of the reports are from people who find a chick and think it is 'abandoned' by its parents. Bev and other hotline operators then explain this is not the case; that they can look after themselves.

Recently, there have been calls from people who are scared of the turkeys. They find them intimidating. One frail elderly lady is too scared now to take her old slow dog outside for a walk because the turkey attacks it. The turkey also stakes out her front door!

Increasingly, calls are from distraught garden owners who want to know whether it is legal to remove the mound and how they can get rid of the turkeys in their garden.

Occasionally they get called out to rescue an adult brush-turkey with an injured leg. The problem is not much can be done to heal the legs, but they can often heal by themselves or the turkey adapts so it can still scratch up food or nest material. Unless it is badly injured and not able to feed itself, it is better to wait and see whether they can cope okay. If badly injured, wildlife rescuers will try to catch it and take it to a vet.

I later found out that Bev forgot to mention that she is a recipient of the Order of Australia medal. Her voluntary work for wildlife spans nearly three decades. Amazing!

FOLLOW UP

Encourage people to seek general information about brush-turkeys (and other wildlife) through their website or email <info@sydneywildlife.org.au>. Bev will then provide the necessary information. Please keep the rescue hotline phone for emergency rescue calls only.

INTERVIEW

JENNY O'MEARA
SYDNEY OLYMPIC PARK AUTHORITY

WHAT IS YOUR ROLE?

66 I am a Parklands Ecologist for Sydney Olympic Park, the site originally formed for the Olympic Games 2000. 99

WHAT ISSUES DO YOU HAVE WITH BRUSH-TURKEYS IN SYDNEY OLYMPIC PARK?

66 Sydney Olympic Park is a suburb in Western Sydney that was originally redeveloped for the Sydney 2000 Olympic Games. It features a large sports and entertainment area but also many residential and commercial buildings as well as extensive parklands. Brush-turkeys returned to these parklands in 2017 and built their first mound in 2020.

So far these birds have not greatly changed the landscape of the park or its ecology, but the parklands are largely man-made and the soils are poor and do not hold moisture. We are concerned the turkeys' raking may cause plants to die because the mulch cover and moisture are reduced, and more weeds may grow instead. The future will tell how many BTs can survive in our parklands and how these affect the existing natural balance between species.

Our park staff are busy spreading all new mulch delivered for landscaping quickly so these birds don't take over the mounds, and they often have to remove leaf litter that is spread across pathways or cycleways and causes a slipping or tripping hazard.

The biggest job we have now is to educate local residents or businesses that know little about these birds and are not aware that they are a protected native species. Some harass the turkeys, others disturb or remove the mounds. For example, one local hotel needed advice when an enterprising male brush-turkey adopted their brand-new, expensive garden at the front door. Our staff now spends considerable time on education about the positive roles these birds play in our environment and about their protected status. 99

4

ARE THEY INTRUDING OR INVADING?

THEY ARE NOT FROM OVERSEAS, THAT'S FOR SURE!

One of the most common remarks in both the media and from those not keen on BTs is that these birds are "*intruders*" or "*invaders*". This makes sense for someone who has lived in the same area for a long time and has never seen a BT in their region before. They have worked hard to establish a tidy, secure dwelling with a neat-looking yard then, one day, mayhem breaks out in tidy-land: a BT moves in and decides to rake together all that tidiness into one big, untidy mound. We have already looked at the types of complaints such an event can evoke (see Chapter 1), but the big question remains: are they *invaders* or *intruders*? Where did they come from?

Did these birds evolve in Australia, or are they an introduced species like the Indian myna or cane toad? We know that the first of these options is correct, with the answer lying far back in history. Fossil bones found in various parts of Australia tell us that the family of birds that the BT belongs to, the megapodes, have been around for at least three million years. Some of today's BT ancestors built mounds so big that they are still visible as fossil mounds in the landscape. We know of five large megapode species that have become extinct over time[1]. Much smaller megapodes also existed, such as a tiny species that only weighed up to 330 grams (11.6 ounces) and was found as a fossil in South Australia[2]. Only three megapode species are left in Australia today: the Australian brush-turkey, malleefowl and orange-footed scrubfowl. All of them are native and not introduced.

Figure 6. Comparison of the size of a BT (left) and its megapode ancestor, which roamed Australia about 1.3 million years ago. Adapted from an article in "The Conversation"[3]

Given that BTs have existed in Australia for aeons, are they *invading* areas where, until recently, they never used to occur, such as backyards in the middle of Brisbane or Sydney? The short answer to this is no: they are not invading. Historical records confirm the extent to which BTs existed across the East Coast before colonialists expanded their settlements, encroached on BT habitats and began hunting these birds *en masse* for food.

To investigate past distributions of BTs, I worked closely with staff from NSW National Parks and Wildlife Service to sift through early historical records mentioning BTs in New South Wales and we published our findings in a scientific paper. We found that while BTs have increased in numbers in coastal areas, they have withdrawn from regions in the southern and western parts of their former distribution. For example, we found that BTs used to occur as far south as near Jindabyne in the Snowy Mountains and as far west as the arid Pilliga region[4].

More recently, Matt Hall and his colleagues from the University of Sydney sifted through historical records from two sources (*The Atlas of Living Australia* and a Citizen Science project) to confirm this shift in distribution. They analysed an astonishing 98,019 records mentioning BTs between 1839 and 2019 and found that the birds moved away from the western and southwestern parts of their former range across New South Wales to expand into north-western areas. For example, by 1939 they had disappeared from the NSW Southwestern Slopes, Southeastern Highlands, and Cobar Peneplain bioregions, and by 1959 from the Mitchell Grass Downs bioregion in Queensland. It is likely that habitat loss, predation and hunting have all contributed to this decline. At the same time as they started to decline in their former natural range, from the 1960s they started recolonising areas where they used to occur, only now these areas were developed and urbanised, had less vegetation and were more densely populated[5].

The change in distribution mentioned above also raises questions about which type of landscape BTs prefer to live in. Their traditional habitat are the closed forests along the eastern seaboard, such as rainforests, wet sclerophyll and even mangrove forests. Where they do venture inland, they are usually found in the wetter ranges. However, these birds are remarkably flexible and can live success-fully in a wide range of other habitats, which is also why they have become the amazing, annoying city birds we now know. They have learned to live in hybrid modern urban landscapes with various natural and man-made terrains and food sources.

One interesting example of the BTs flexibility is its relationship with the prickly pear cactus (*Opuntia stricta*). BTs love to eat the fruit of this introduced cactus, which became a pest species in the 1920s but by then had covered 23 million hectares of New South Wales and Queensland. Half of this infested area was so densely covered with cacti that it was useless for agriculture and abandoned by landowners. The BTs followed the expansion of the cactus distribution and were suddenly found in areas where they had not been seen before. When the government introduced a moth to kill the cacti and help the farmers, the BTs retreated again from most of those areas where the cactus was no longer found.

Today, BTs range down Australia's East Coast from Cape York in the tropics to the Illawarra region south of Sydney. Occasionally they are sighted further south, all the way to Eurobodalla in New South Wales. But how far inland can they be found? In the southern part of their range, such as Sydney and Wollongong, they only venture up to 100 kilometres (62 miles) west, such as to the western boundary of the Blue Mountains National Park. Further north, they can be found as far as 500 kilometres (310 miles) inland from the coast, predominantly in areas that are forested, such as large national parks or state forests. BTs have also been introduced to Kangaroo Island in South Australia, though that population may have suffered considerably from the damaging bushfires in 2020.

Figure 7. Distribution of the Australian brush-turkey. Data from the Atlas of Living Australia[6]

We don't know precisely each small area where BTs have occurred in the past – the records are not accurate enough for that sort of information. But there is a solid indication that these birds are *recolonising* areas where they used to occur, rather than *invading* or *intruding*. This sentiment is widely accepted and has even been included in a recent Channel Nine news feature[7]. Chapter 5 looks at why so many are returning now.

Why Are They Returning Now?

Suburbs as Brush-turkey Heaven

One of the main reasons why BTs disappeared for so many decades is that people used to hunt them for food, especially during the Great Depression in the 1930s. The birds were nearly wiped out because they were so easy to track down. At that time, when food and income were scarce, people welcomed the BTs' extra meat and large eggs, which are bigger than goose eggs and contain a lot of yolk. It was common to dig up the eggs as, unlike today, they did not enjoy any legal protection. The Country Women's Association provided recipes for BT egg omelette, and it was well known that the meat should be hung for a day to be less tough to eat.

But the hunting did not start in the 1930s. Even in the 1860s, less than 80 years after British settlement, the BTs had already become scarce in Sydney, as evident from a *Sydney Mail* article from 1868 that comments on the BTs scarcity and provides the following reason: "Its flesh is exceedingly good, and its eggs are reckoned a delicacy"[1]. Since 1974, Australian native animals have enjoyed legal protections under the National Parks and Wildlife Act 1974, as well as under the Threatened Species Conservation Act 1995 and the New South Wales Biodiversity Conservation Act 2016. Under the Biodiversity Conservation Act, killing BTs or collecting their eggs is illegal and attracts heavy fines. It appears those few BTs that survived in small pockets across New South Wales have made the best of new legal protections. Our current generation of native BTs no longer perceive humans as a significant threat.

Game.

and add with pepper and salt to taste, add the wine and juice of a lemon also and pour over the pigeons, allow them to remain in the oven for about a quarter of an hour. Serve with mashed potatoes.

437—ROAST SCRUB TURKEY.

1 turkey	Piece of bacon fat
Little flour	Dripping.

MODE.—Pluck and clean the turkey nicely, rub it over with a little flour, put it in a baking tin with the dripping, place pieces of bacon fat over the breast, keep basting it well all the time and bake for an hour. Serve with bread sauce.

438—THE SCRUB TURKEY.

The scrub turkey is a very small bird, not much larger than a wild duck, with a breast like a pheasant and flesh as white, in fact, I have often served it as pheasant and people have not known the difference. It is a most delicious bird, one of Australia's best

439—ROAST WONGA PIGEON.

6 pigeons	Juice of 3 lemons
Saltspoonful of cayenne	½ a pound of butter
Teaspoonful of salt	2 cupfuls of breadcrumbs
	1 tablespoonful of chopped parsley.

MODE.—Pluck and clean the pigeon nicely, rub it over well with flour, pepper and salt them well, make a stuffing with half the butter, all of the breadcrumbs, chopped parsley and pepper and salt, divide it into six and equally stuff each bird. Then squeeze the lemons into a basin and beat up the butter with the juice until it is like a cream. Place the pigeons in a baking tin and cover each one well with the lemon and butter, place in a smart oven and bake for half or three-quarters of an hour, baste them as often as you can. Serve at once with watercress (if obtainable).

Figure 8. Page from a cookbook from 1903 describing BTs as "a most delicious bird, one of Australia's finest"[2]

A second reason for their recent comeback is that introduced foxes have, in the past, considerably reduced the numbers of BTs and many other native birds. The baiting of foxes with poison has increased over the past few decades, meaning all native birds have a reprieve from these predators. BT chicks and adults were easy prey for foxes and now they are one of many native species to benefit from reduced fox numbers. However, the war against foxes is not over. They continue to spread throughout our towns and cities and are much more common than you may think.

When discussing the BTs' return, we need to consider where they like to live. Their chicks survive much better in thickets (see Chapter 13), and Australia's East Coast is riddled with introduced weeds like lantana that cover large areas and are impenetrable to humans. We introduced these weeds either purposely or accidentally as garden plants that then escaped to thrive in the wild, and they are now a good hiding spot for BT chicks. For adult BTs, suburban gardens are often the oasis they are seeking. In the past, many people loved their overly manicured English garden, with a lawn, hedge and three rosebushes, but now these are increasingly replaced with more natural-looking gardens that are bushy and filled with mulch and native plants. Such backyard refuges are attractive to BTs as they resemble their natural habitat: the closed forests, especially rainforest. In addition, gardens often come with a food supply in the form of bird feeders, pet food or moist mulch rich in worms and other creepy crawlies that BTs like to eat. Water can be found in bird baths, and the males eager to build a mound find mulch and leaf litter as building material.

Finally, we also have to consider how the environment has changed from a BT's perspective. All along the coast, we have successfully cleared many of the forests that resemble their natural habitat. Having lost their original place to live in, these birds are now adapting to the suburban environment we created. In some ways, this is really a big evolutionary experiment to test how adaptable BTs can be. In fact, Dieter Hochuli from the University of Sydney points out that cities are becoming increasingly important as a refuge for various species of native wildlife, especially during bushfires and extreme weather events[3]. The BT is one of those birds that benefits from the city's protection when their natural habitat is either logged, burning or drying out.

MEETING

TEAMS MEETING WITH: **Lee De Gail, Ranger, Sydney North Area, Greater Sydney Branch, NSW National Parks and Wildlife Service**

SUBJECT: **Why are Brush-turkeys (BTs) returning in large numbers?**

DATE: **27.10.2023**

NOTES:

He is unaware of many BTs in the Northern Beaches area before approximately 2013. After that, the population increased quickly. Lee thinks this is partly because of the increased fox control that NPWS has been conducting. The new rules for dog containment also helped the birds, as dogs do not roam around as freely as before.

Most homes in the Northern Beaches area come with beautifully landscaped gardens. The same applies to nearby areas, such as Wahroonga, Chatswood, and Northbridge. In all these gardens, you find nice mulch and big trees that drop lots of leaf litter, offering an ideal location for BT feeding and mound building.

Hunting decimated these birds in the past, and he thinks very few, if any, survived in Sydney during the Great Depression. Nowadays, people are adapting to living with BTs and the heavy fine that comes with killing native wildlife, including BTs.

Lee sees the BT as a huge story of comeback. A bird that has been hunted almost to extinction has been heavily affected by hunting and the reduction of previous habitat and is now adapting so well to life in the suburbs.

One recent study examined how far individual BTs can migrate within a large city like Sydney. Matthew Hall, John Martin, Alicia Burns and Dieter Hochuli from the University of Sydney marked individual BTs with wing tags. They found that some of these BTs travelled between 8 and 37 kilometres (5 and 23 miles) from where they had been tagged — without returning[4]. From my own research, we also know that the chicks, which can fly much better than the adults, regularly disperse between 100 to 220 metres (109 to 240 yards) from the mound during the first days after emerging, sometimes even up to 800 metres (875 yards)[5]. Therefore these birds are highly capable of colonising new areas and we can expect them to keep moving south as long as the climate remains suitable for maintaining the proper incubation temperatures in their mounds (see Chapter 8).

When we study the presence of BTs in certain areas, we often rely on members of the public (known as 'citizen scientists') to enter their observations into either a database or an app, like the Big City Bird App (see Chapter 18). While most people know what a BT looks like, the odd member of the public still mistakes them for other birds. So, what are the identifying features of BTs, and what are some of the special things worth noticing about their appearance?

Section 2
The Amazing Brush-turkey

CHAPTERS 6-14

Top: Brush-turkey hatchling, one day old, next to brush-turkey egg.
Photo: A. Göth

Left: Male brush-turkey.
Photo: E. Mann.

6

WHAT DO THEY LOOK LIKE?

THE SECRETS OF UV REFLECTION AND VERTICAL TAILS

BTs are approximately the size of a domestic hen turkey, or for Australians, the size of a bin chicken (aka white ibis). BTs have a body weight of approximately 2.2 kilograms (4.9 pounds) for females and 2.5 kilograms (5.5 pounds) for males, and their length is about 60 to 75 centimetres (23.5 to 29.5 inches). When they spread their wings, their wingspan is about 85 centimetres (33 inches).

Their predominantly black or brownish-black feathers provide good camouflage in the dull light of a rainforest. Only their chest feathers have a greyish-white fringe. The red head and yellow neck wattle are largely bare, except for a wispy layer of short black stubbly feathers on the head. Their legs and long claws are brown, grey or greenish-yellow and have a muscular build. If you have noticed those long toes and nails, well done — this is why they are called megapodes! Derived from Latin, *megapodes* means 'big feet'. It is the dominant characteristic for 22 species of megapodes, all of which spend a large part of their life digging either in nesting sites for incubation or in groundcover for food[1].

Males and females are similar in size, though an experienced observer can tell that the male is slightly taller and has a sleeker appearance. His head usually also appears a brighter red, whereas the red of the female's head is duller and covered by more black feathers than in the male. The yellow throat wattle is always small in females, whereas it changes size in males during the breeding season (see below).

Male & female: telling them apart

Red head with some black feathers

Male brush-turkey

- Slightly longer legs
- Sleeker appearance
- Wattle bright yellow
- Wattle longer in breeding season
- Less black feathers on head in breeding season

Ear
Yellow wattle
Two-layered tail

Female brush-turkey

- Plumper appearance
- Wattle duller yellow
- Wattle small ruffle only
- More black feathers on head

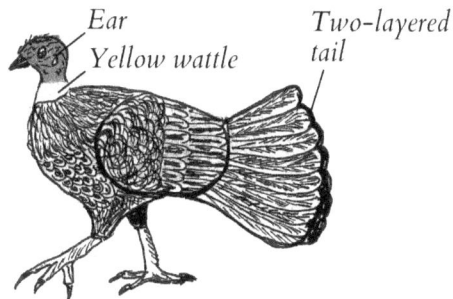

Figure 9. Comparison of male and female brush-turkeys.

All megapodes, including BTs, belong to the order of birds called Galliformes. This order dates back 30 million years and includes chickens, peacocks, pheasants and quails, which explains some similarities in looks. Within the Galliformes, the family of megapodes are the most ancient members. Scientists call them a sister group of all other Galliformes birds that exist today[2].

While the adult BTs are not exactly cuddly looking, the chicks definitely are. At hatching, they weigh between 100 and 170 grams (3.5 to 6 ounces), are brown and fluffy, well-camouflaged and look rather round as they don't yet have a prominent tail. Unlike their parents, they do not have a bald head, making them look less vulture-like, though they already possess the large feet typical of megapodes.

'Cute' is not something that BT chicks can be called for long, as they grow up in a slight hurry. After only a month, they have already grown a short black tail and start developing black feathers on their body. Within two to three months, they resemble small adults without any of their fluffy brown baby feathers left. The red skin on the head is visible in young birds but partly covered by stubby black feathers for many months. Only adult birds, especially the dominant males in BT society, showcase a bald red head with only a few black feathers near the beak. Within nine to ten months, BT chicks weigh ten times as much as when they hatched.

Growing up

1 year old

- Full adult size
- All black
- Bright red on head
- Bright yellow wattle in males
- Duller wattle in females

8 weeks old

- Mostly black
- Head more red
- First yellow on neck
- Tail feathers longer

ear

4 weeks old

- First black feathers
- Tail feathers 1/3 of body length
- More red around eyes & ear

2 days old

- Mostly brown
- No tail feathers
- Fluffy feathers except on wing

Drawings not to full scale

Figure 10. The growth of chicks from hatchling to adult size.

The tail starts to show at about three weeks of age and is a remarkable feature for those who know about birds: it spreads out vertically, unlike the horizontal tails found in most other birds. There are two benefits of such a tail orientation. First, a vertical tail is less likely to get dirty when the bird vigorously kicks soil and leaf litter backwards during foraging and mound building. Second, a vertical tail makes it much easier to navigate through obstructing vegetation in the natural habitat of these birds, the rainforest.

When navigating dense vegetation, BTs fold their tailfeathers together like a closed fan. At other times, when they strut around in the open, some of them fan their tail feathers out wide, possibly to show off who is more dominant in BT society and who is not. This spread-out tail looks rather impressive, mainly because the tail consists of two layers of feathers which, when spread out, form an almost petal-like fan.

How will I wear my tail today?

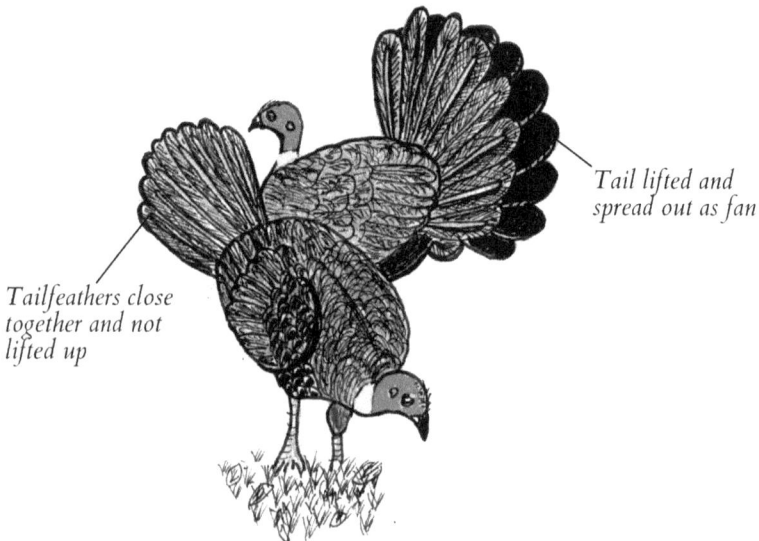

Tail lifted and
spread out as fan

Tailfeathers close
together and not
lifted up

Figure 11. The two ways Brush-turkeys wear their tail.

There is one more fun fact about the tail, even though it is no fun for the turkeys. When predators such as dingoes or ill-intentioned humans try to grab them by the tail, BTs can shed all their tail feathers spontaneously[3]. Like a lizard sheds its tail to escape a predator, BTs do the same! I have seen a BT without tail feathers more than once, and I must admit they look much less mighty!

Another dominant feature in these birds is the throat wattle, a yellow fleshy lobe around the base of its neck. Some people call it a collar or neck pouch. In females and males during the non-breeding season, the wattle resembles a wrinkled area of yellow bare flesh. But when those hormones kick in ready for reproduction around July each year, the male's wattle becomes longer and droops into an elongated loose pouch that swings back and forth, sometimes even reaching the ground. It is not a permanent structure, though. Surprisingly, the males can retract and extend the wattle quickly. I have caught many adult males with impressively large wattles only to find these promptly retracted into a small ruffle.

Where's the wattle gone?

5 minutes later: wattle retracted

Male with wattle extended

Figure 12. Males can extend or retract their wattle quickly.

We don't know how the males control the size of their wattles, but it may have something to do with pumping air or blood into the fleshy structure. We know that in the wild the males use their wattles as signals to other males and females during social interactions. Dominant males fully extend their wattles while subordinate ones retract them when the dominant ones are around[4].

In addition, there might even be a signal in wattles that is hidden from us humans. We know that birds can see UV light, but we can't. One scientist from the University of Queensland, Valerie Olsen, held a UV sensor against BT wattles. She found that they shone brightly under the UV, much brighter than in the normal light visible to humans. It may well be that the UV reflection off wattles signals dominance and hierarchy to other BTs even though we humans can't see this because it is outside our vision spectrum[5].

On the topic of wattles, you may like to know that some BTs have lilac- or mauve-coloured wattles instead of yellow ones. The only problem is that to see them you have to travel to the northern tip of Australia to the Cape York area. The BTs there are a different subspecies (*Alectura lathami purpureicollis*) and only differ from the other BTs in the colour of their wattle.

If you like odd things, you may also like to travel to Noosa or Townsville in Queensland. Here, a genetic mutation amongst the local BT populations occasionally leads to a white BT (scientifically they are not albino, but called 'leucistic'). These individuals have white bodies with a yellow wattle and pale red head, and they quickly become famous on social media as they look very different from their black relatives[6]. One particular white BT in Noosa became famous among the tourists and locals, strutting through the main tourist area and holding up traffic while he built his mound. He became so popular that locals mourned his death in a car collision. They even approached the council to erect a plaque and burial site for the much-loved bird.

When trying to distinguish males from females, the throat wattle is the best tell-tale sign for males. In the non-breeding season, when wattles are smaller, this separation becomes more challenging and you need to look for the differences shown in Figure 9 above. Those familiar with handling birds (under an appropriate license, of course) may also inspect the cloaca. Both sexes have a phallic structure on the lip of the cloaca, but in males this is considerably larger and typically darker pink[7].

Speaking of the non-breeding season, this is also when you are more likely to see a bedraggled-looking BT with some tail or wing feathers missing. Like all birds, BTs need to replace their feathers annually. We call this moulting, and you can read more about it in Chapter 9. One thing is for sure: by the time the mound-building season comes around, they will be looking their best again! But why do they build a mound at all?

7

WHY THE MOUND?

A HIGHLY UNIQUE INCUBATOR

D o you think you are working hard? You might not be the only one: imagine shifting dirt 22 times the weight of your own body every day for a few weeks while you construct a mound. For a 2.5-kilogram (5.5-pound) BT male, this means shifting about 56 kilograms (124 pounds) daily to create a mound containing 2 to 4 tonnes (2.2 to 4.4 tons) of material. For an 80-kilogram (176-pound) man, this equates to moving 1,760 kilograms (3,880 pounds) of dirt daily, and a 50-kilogam (110-pound) woman must deal with 1,100 kilograms (2,425 pounds). Now you know how hard a BT male works when he creates a mound that will be his nest.

Darryl Jones measured some mounds in Queensland and found that, on average, they contained up to 4 tonnes (4.4 tons) of soil, leaf litter, sticks and plant materials. They measured up to 1.3 metres (4.2 feet) in height and about 4 metres (13 feet) across, roughly the size of an average car[1]. In New South Wales, I have seen mounds even bigger than that, up to 1.5 metres (5 feet) high and 5 metres (20 feet) in diameter. And in an even colder climate, on Kangaroo Island in South Australia (where BTs are an introduced species), mounds contained an average of 6.8 tonnes of material and were roughly a metre wider than those measured by Darryl in Queensland. This makes sense if you think of heat as a driving factor behind the decomposition of the organic material. Mounds can be smaller in warm, tropical environments such as Queensland, but need to be larger in southern climates where it is colder outside[2].

Size comparison

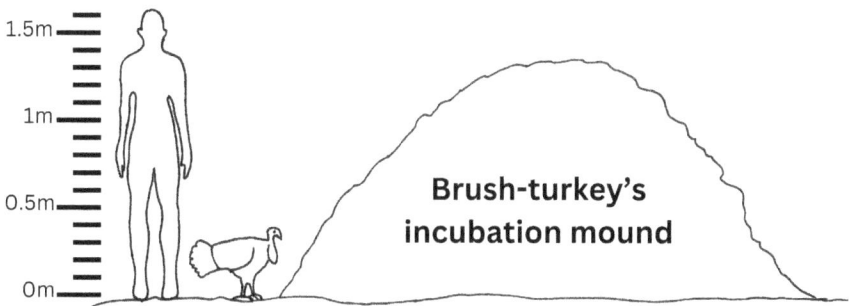

Figure 13. Relative size of a brush-turkey compared to its incubation mound and an average-sized person.

A mound as a nest? This is highly unusual and not found in any other group of birds except the BT and its megapode relatives such as the malleefowl and the orange-footed scrubfowl. Bird nests provide a temperature-controlled environment for eggs to incubate, and in typical bird nests the parent birds use their body-warmth to keep the eggs warm. But BTs and other megapodes have developed a different temperature-controlled environment to incubate their eggs. In BT mounds, all that plant material starts to break down and rot, just like in a big compost heap. This rotting, scientifically called 'decomposition', is caused by microorganisms and fungi, and it produces heat for incubation. Males want only plant materials in their mound, but suburban mound-builders might end up incorporating all kinds of other stuff, such as soccer balls, sprinkler systems, lost teddy bears or even a wedding ring. These are just accidental inclusions because they happen to be in the male's path while he builds his nest.

So why all the effort if other birds are just content with building a small nest, laying their eggs and sitting on them to incubate? Let's call it an alternative lifestyle that evolved millions of years ago and freed BTs and other megapodes from several parental burdens. First, instead of being stuck in a nest sitting on eggs for incubation, they can roam more freely – the warm mound does the incubating for them. Second, instead of laying only as many eggs as a small nest can fit, megapodes can lay more – there's plenty of space in a BT mound!

If it's such a great deal for the parents, why don't more birds do it? This alternative approach can only work in certain climates where the male is able to keep the mound temperature stable at around 32° C to 34°° C (89.6° F to 93.2° F). Obviously, this would be more difficult to achieve in colder climates. It explains why mound-building megapodes are only found in tropical climates or regions where it is at least warm enough to maintain suitable temperatures in their mounds during certain parts of the year. It is also why the BT has a specific breeding season (see Chapter 9).

There are several other reasons why 'real estate BT-style' has not taken off as the most popular investment amongst birds. For one thing, the male needs to put considerable effort into the mound – not only into building it but also in maintaining its stable temperature (see Chapter 8). Plus the success of this method is highly dependent on two environmental factors: temperature and rainfall. It only works well when outside temperatures are not too cold or hot, and there needs to be enough rainfall for the microorganisms and fungi to do their job of decomposition and heat production (see Chapter 8). In other words, BTs have a higher chance of failure when these conditions are not met.

There are three more reasons why mounds are not the preferred style for other birds. One is that the microorganisms in the dirt can be fatal to the eggs, causing them to rot. While BT eggs have some remarkable features to protect them against such deadly intrusions (see Chapter 14), egg mortality is still comparatively high: about 14% of eggs don't turn into chicks[3]. Also, chick mortality is higher than in most other birds partly because their young receive no parental care at all (see Chapter 13). Lastly, the unusual compost-style incubation of BT eggs brings an additional burden to the females: they must regularly develop an enormous egg to produce a highly developed chick with some chance of survival. Let's assume you were a human female of 50 kilograms (110 pounds) and you had to produce a 5-kilogram (11-pound) baby every two to five days. You may now have a different view of BT females being lazy and just roaming around: they spend all their time searching for enough food to create their gigantic eggs. Once they've laid their egg, they let the male stick around to maintain and defend the mound as they are busy producing the next precocial offspring.

If so few birds use mounds for incubation, you may ask, how did it all start? Did it evolve as a separate approach while other birds started building nests, or is it an even more ancient trait that evolved long before other birds started building nests? None of these assumptions is correct. Genetic testing (DNA sequencing) of megapodes and other birds strongly suggests that the ancestors of the BT and his megapode relatives were birds that built more conventional nests like those used by ducks or chickens, a theory also supported by the temporary presence of an egg tooth during modern BT chicks' gestational development (see Chapter 12). It would appear that only later in evolutionary history did the megapodes abandon traditional nest building and incubation methods but the exact reason why remains unknown[4].

Now we know the purpose of the mound, how does the male BT build a car-sized incubator of leaf litter all by himself?

8

MOUND-BUILDING - HOW DO THEY DO IT?

CONSTRUCTING A MOUND IS NO EASY FEAT!

If you had to use only one word to describe how a BT builds his mound, you would probably choose 'obsessively'. Maybe 'single-mindedly' would also do. Once he starts, there is not much that can stop him. Every year, sometime between May and August (see Chapter 9), the male's first big decision is where to build his mound. Even though we may regard those locations as random (and often highly annoying), he is actually fussy about where he will establish his real estate. He usually likes a certain degree of shade above the mound and, apparently, the quality and quantity of available leaf litter nearby also affects his decision[1]. Given that a certain degree of moisture is necessary for the microbes to decompose the leaf litter, we can also assume that locations in wet gullies, at the edges of lakes or creeks or under large shade-providing trees are more suitable than those in open, drier country. However, too much moisture is not good either. Physics tells us that a higher water content means that heat moves away (conducts) from the mound more quickly. Also, the embryos in the egg cannot cope with too much water content in the mound. It will stop them from breathing properly because wet mounds are less permeable for carbon dioxide and oxygen[2]. So, the male's choice of mound location will determine his reproductive success.

Considering this, it is increasingly obvious why a nicely mulched garden, slightly moist from sprinklers and shaded by some trees, is a prime building site from a BT's perspective. We don't know whether he also chooses a site based on how many thickets there are for the chicks to hide in, but you will notice that quite a few mounds are found near dense undergrowth. We know that chicks survive better if they have thickets to hide under (see Chapter 13), but cannot ask a BT whether he considers the future survival chances of his kids when choosing his nesting site.

While some males, especially young ones and those without a previous territory, will start building a new mound from scratch, others will make it easier for themselves and build on top of a mound from last year. Why build the foundations again when the brickwork has already been laid? Even though this old material is mostly fully decomposed, it may still add a bit of warmth when turned over. It also adds some insulation between the eggs and it helps to make the mound appear higher. There is a strong indication that females choose males partly based on how impressive their mound looks[3].

Once he has chosen a site, the male does not just rake together plant material randomly. He has a plan. Starting as far as 100 metres (109 yards) away from the mound location, he rakes together the material into a narrow strip to create something akin to a red carpet that leads to the mound, except that the carpet is made of leaves, sticks, twigs, soil and, in the city, all the odd bits he finds in a park or garden. This process takes him a few days. Only when this leaf carpet is complete does he start the real work of scratching it all towards the mound, and he works tirelessly. If you live near one, you can hear him rake almost all day, except for short feeding or roosting breaks or when he needs to chase away other males and even females. He tends to be very defensive of his new mound and needs to make sure another male does not take over and benefit from the work he has already done.

Building a mound

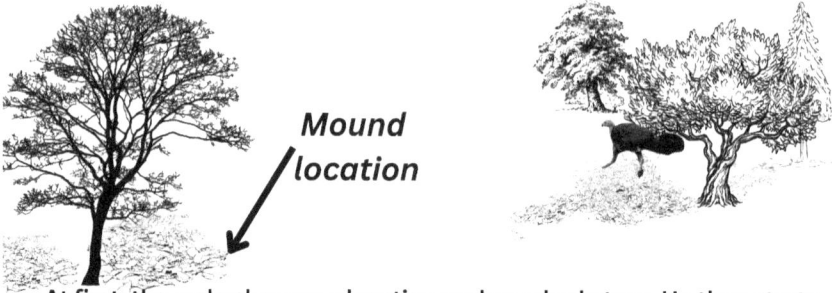

Mound location

At first, the male chooses a location under a shady tree. He then starts raking together leaf litter up to 100 m away from the mound.

Using that material from far away, he builds a trail of leaf litter that is heading toward the mound location. This takes several days.

Only in the last stage does he scratch it all together into a mound. The floor in the area nearby is now raked clean.

Figure 14. The different phases of mound-building.

It is no exaggeration to say that the male is utterly obsessed with his mound and not easily distracted. In the city, this leads to new challenges. I have witnessed a male building his mound across a road, completely undeterred by cars approaching. The cars had to stop until he had raked all that leaf litter to his mound site on the other side of the road. Recently, I was sent a video of the most unusual kind: a male raking his mound material out of a fire pit with smouldering fire and smoke almost touching his feet. This happened because the garden owner had raked up all his leaf litter in one big pile a few weeks prior, right next to the fire pit, ready for a burn later on. The BT had taken over this pile as his mound, and when the owner started shovelling the leaf litter into the fire pit, the male was determined to rake it back to where it obviously belonged!

The fastidious maintenance phase begins once the male BT's building efforts result in an actual mound. He will still rake some material onto the mound daily, but now his main job is ensuring the temperature inside becomes stable enough for eggs to be incubated. If he fails, females will not trust him with their eggs, and he will not get to mate with them either (see Chapter 10). Just like teenagers must learn new skills, young BT males must learn how to build a good incubator, and it appears this may take a while. In my aviaries, I had young males start building mounds from the ages of four-and-a-half to nine months[4]. However, as Darryl Jones observed, such youngsters are not often successful in attracting girls (see Chapter 8).

The decomposition inside the mound leads to a stable temperature after about six weeks. At the start of mound-building, when those fungi and microbes first start feeding, the temperature can skyrocket to almost 40° C (104° F), much too hot for the eggs[5]. While the male depends on the work of these tiny organisms, he must also provide them with either more or less plant material to decompose, meaning he must know the temperature inside the mound so he can decide how much of this material to provide. The females, too, must be able to test the temperature when they decide where to lay their eggs (see Chapter 10).

So, how do BTs measure the temperature? They must have some kind of sensor in their body to do so, and scientists believe this structure is most likely found inside the beak[6]. Nobody has yet looked at the inner structure of a BT beak under the microscope to confirm this assumption, but there are several reasons why it is the most likely location for this sensor. First, BT males often take small amounts of soil into their beak while working on the mound's temperature regulation and females also do the same when digging holes into the mound. Second, domestic chickens definitely have a temperature sensor inside their upper beak[7] and BTs belong to the same bird order as chickens; scientists often infer that if a structure is found in one species, it may also be found in a closely related species due to their shared evolutionary history. Some other related megapode birds, such as the malleefowl, also take soil into their beak while they are regulating the temperature of their mound[8].

Every day the male digs a little tunnel towards the mound's core, where the eggs will be laid, and takes some decomposed plant material into his beak. This most likely tells him whether the core temperature is close or far from the desired 33-34°C (91.4° F) that the eggs prefer for incubation. Once the testing is done, he springs into action. If the mound is too hot, he disperses some of the plant material from the top and spreads it out to release some heat from the core. You will then see a shallow depression on top of the mound. If he finds it too cold, he rakes more material from the mound's base to the top to provide additional insulation. This creates a more cone-shaped structure. Scientists found that adding as little as 1 centimetre (0.3 inches) of leaf litter to a mound can raise its core temperature by 1.5° C (34.7° F)![9] Hence a BT mound is not just a bit of material raked together randomly: it is an incubator that requires daily tenacious housekeeping. The male cares not only for the eggs inside but also for the microbes and fungi that need to be regularly fed new plant material to keep producing heat.

The males not only maintain the temperature, they also vigorously defend their mound against two potential enemies. First, against other males who are too lazy to build their own mound and instead rely on brute force to take over one already made. Second, the males also defend their mound against predators who may be after the eggs inside. This includes goannas, dingoes and even snakes or wild pigs that want to have eggs for breakfast. The male will try to attack these uninvited guests by running towards and pecking at them and, if they have a tail, he will attack that part first. The pecking-the-tail method is so successful that males often chase away goannas and even snakes much larger than themselves. Some goannas, though, are not deterred and dig into the mound to steal an egg. In that case, the male vigorously kicks dirt at them, covering them almost completely until the reptiles are ready to leave.

All this is a lot of work for the male, so when does he stop? When is his mound large enough? There are pros and cons for having a large mound, especially one that measures 3 metres (10 feet) in diameter and about 1.5 metres (5 feet) in height. On the positive side, large mounds provide a better buffer for changes in the outside temperature, especially during droughts or heavy rainfalls. Large mounds may also attract the ladies (see Chapter 10). On the other hand, large mounds mean more digging and physical exertion as the male mixes the material and digs tunnels during his daily maintenance routine. They may also be more challenging to keep at an even temperature, leading to significant variations in incubation temperatures across different parts of the mound (see Chapter 15). The females, too, must dig deeper holes in large mounds to reach the core during egg-laying.

Considering these pros and cons, a mound should have an ideal midpoint size. However, there is another factor that the males often have little control over, especially in the city: the availability of suitable organic material for decomposition. In a rainforest, they are guaranteed to have continuous leaf litter falling to the ground from the surrounding trees. In the city, they often have to put up with less suitable building materials. Gravel, branches, twigs, small logs and various debris don't decompose much, nor do sprinkler systems, garden hoses, solar lights or garden gnomes. Similarly, eucalyptus leaves and casuarina needles decompose at a much slower rate than other leaves, and these types of trees are common in the suburbs. With less suitable material around, BTs must build a much bigger mound to provide enough sufficient material for the microbes and fungi to decompose.

Once a mound is established and maintaining a suitable temperature, the females will hopefully become interested, usually about six weeks after the male started building. However, not all males are successful in attracting the girls: those with the best incubation temperatures in their mounds receive more and also larger eggs from the females[10].

The breeding season nears its end in mid summer, usually around late January in New South Wales, but it also depends on how much rain has fallen. In Queensland, it may be a bit later. There will still be chicks hatching for a few more weeks, as the eggs incubate for about 49 days, but females will have stopped laying eggs. Once all activity on the mound has ceased and the male has abandoned it, seedlings may start sprouting out of the mound and the floor nearby will be covered in leaf litter again. At this point, once the mound is no longer incubating eggs, it can be legally and safely removed (see Chapter 17). The decomposed soil inside, up to 4 tonnes (4.4 tons) of it, makes excellent compost that you can spread onto your garden if you wish.

We have now explored the routines around mound-building, but there is more to discover about the BT's annual and daily routines both on the mounds and away from them.

9

BRUSH-TURKEY'S CALENDAR AND ROUTINE

WHEN TO EXPECT MOUNDS, CHICKS AND OTHER TURKEY ANTICS

BTs don't care about the months on our calendar. For these turkeys, the big event is not Christmas but the availability of enough moisture after rainfall. Rain in the winter months appears to trigger males to start building their mounds. It will take a few months for the temperature in their real estate to become stable enough for incubation, so they start building while it's still cold to ensure a warm-enough nest by spring when the outside temperatures start to rise. It also means that the breeding season starts later if there is not much rain during the Australian winter.

For those unfamiliar with the climate on Australia's East Coast, we have four seasons in the sub-tropics (southern part of the BT distribution) and two seasons — a wet and a dry season — in the far northern tropics. It does get colder in winter, though this difference becomes less pronounced the further north you move. Accordingly, in Queensland BTs build their mounds earlier than in the southern state of New South Wales. We don't even know when this usually happens in Far North Queensland, and I would be delighted if some of my readers could let me know. But in southern Queensland, mound building follows the first decent rainfall (above 100 millimetres, or 4 inches) in May or June[1]. In Sydney, mound building usually doesn't start before July or August.

Once the building has commenced, the females spend most of their time searching for enough food to produce their enormous eggs (see Chapter 14) and checking out mounds to choose the best nests for their offspring. The males who successfully built or stole mounds are preoccupied with maintaining the temperature and defending the mound. They don't venture far from their nest during the day, but in the evening they join the females and even other males in the trees.

High in the treetops is where they all roost for the night, usually in large groups, and up to 20 metres (65 feet) above the ground! They don't snuggle up like other birds and instead keep their distance, but they choose the same tree as a communal roosting site. It may well be that they do so to have safety in numbers, as they have a lower chance of being picked out of many when a large predator, such as a powerful owl, decides to hunt. The need to avoid aerial predators at night might also explain why BTs are sometimes seen perched on the very ends of flimsy branches, right among the leaf cover, instead of the thicker branches where less leaf cover is available[2].

Young males and those who are not at the top of the pecking order may not be successful in establishing a compost incubator and have a different routine. They roam around independently, trying to find a new site to build or waiting for their turn to move up in the BT hierarchy.

Brush-turkey Calendar

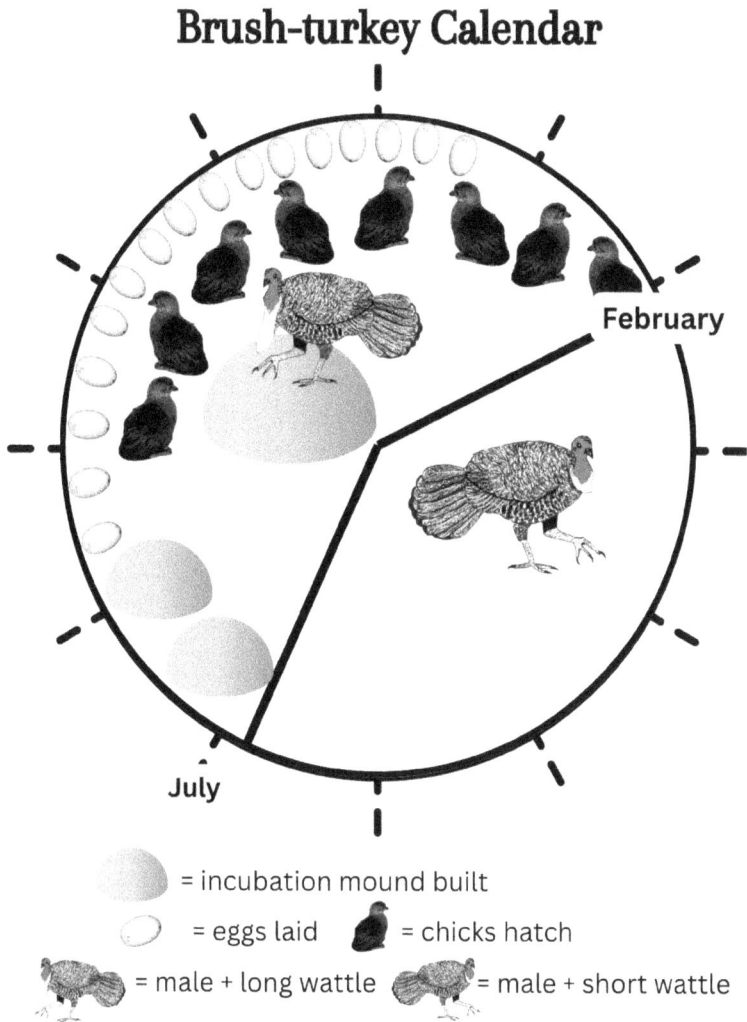

February

July

= incubation mound built

= eggs laid   = chicks hatch

= male + long wattle   = male + short wattle

Figure 15. The brush-turkey's annual calendar.

Everywhere, even in Far North Queensland, BTs stop breeding sometime around the end of January after the incubation mound has produced chicks for up to eight months. BTs now need to invest their energy into another important life-saving activity: getting changed. In birds, we call this moulting, meaning they replace their feathers. Old feathers may have become worn from digging in the dirt, fighting with each other, feather mites eating them and just general wear and tear. Feathers are crucial for survival. Firstly, they enable flight, even though adult BTs prefer to run rather than fly and only do so when they need to escape a predator or venture up a tree. Secondly, feathers help the birds stay warm and keep dry. Their black feathers also make them blend in with their environment to give them camouflage protection from predators — at least in their natural habitat, the rainforest. Finally, nice-looking feathers that are shiny and not worn at the edges give the male a better appearance, which may affect how often the girls choose him and his mound for egg laying. We don't know enough about this aspect of BT mate choice but, in most birds, a healthy-looking plumage attracts more females than a drab one. For all these reasons, BTs lose a few feathers at a time, regrow those then lose the next ones until they have changed their entire outfit ready for the next breeding season.

So, the adults' year is divided into a breeding and non-breeding moulting season. What about the chicks? Chapter 13 describes their unusual upbringing, but you can expect to start seeing them about seven weeks after females have laid their first eggs. In Sydney, that may be in November, whereas further north, it will be earlier in the year. December is often peak hatching time, and if you have a mound in your garden you may be lucky enough to spot one of those shy little hatchlings making its way into the world.

Other birds include a dedicated time for singing in their daily routines, often around dawn and dusk. BTs do not sing. They are not songbirds as they are missing the sophisticated syrinx that enables songbirds to sing. All that BTs manage is a few sounds, but they are generally a quiet bird and are heard only occasionally throughout the day. If you watch a male near his mound, you will most often hear a deep booming call that he produces when he inflates his wattle with air then forces that air through his nasal openings (nostrils). This creates a low-frequency booming noise with either one ('oom') or two syllables ('oo-oom') or sometimes more. The males mainly utter this call when their rivals are nearby or when they want to attract females to the mound. In the presence of a girl, they may add softer and very deep grunts, which she may then repeat. These grunts are sometimes also heard from females and males feeding together in one area. They sound like a deep 'bock bock', a bit like the sound of a bass drum. You can listen to the calls online[3].

The male's boom sound

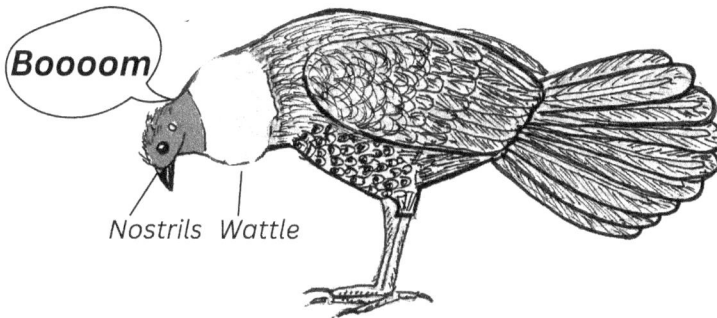

Boooom

Nostrils Wattle

Air from the wattle is exhaled noisily through the nostrils.

Figure 16. A male brush-turkey uttering his typical "booming" call.

While calling is not a regular part of their daily routine, cleaning their plumage definitely is. Like all birds, BTs spend considerable time preening their feathers with their beak throughout the day. They stroke every feather from its base to the tip to remove parasites and dirt and align it in its optimal position. Apart from that, they take sand baths, also called dust baths. Land-fowl often don't bathe in water but use sand instead. They dig a slight depression in the ground where the sand or soil is dry then wallow in it, flapping their wings frantically to create a cloud of dust that covers all their feathers. This dust absorbs excess oil from the feathers, stops them from becoming greasy or matted, and smothers tiny organisms that feed on feathers such as lice and feather mites.

Sand bathing is not the only bath BTs enjoy. On sunny days, you can see them lying on their side with their wings spread out, and many people think they are witnessing a dying or injured BT. Instead, this is entirely normal behaviour and is called sunbathing. Like dust, the radiation of the sun removes nasty parasites from the feathers[4].

Now that we have an idea of the BT's annual calendar and daily routines, we should consider how many years a BT undergoes this cycle. We don't know how long they live in the wild, but we have some indications that their life span might be around nine years, or more. Three BTs that carried leg bands were seen for nine consecutive years in a patch of rainforest, and one male that was recognisable by his bent beak lived on the same property for seven years before he disappeared[5]. If you have any information on the longevity of individual BTs that you can clearly recognise, please let me know!

10

HAVE A MOUND? GOTTA WATCH IT!

THE PERFECT BOREDOM BUSTER

Has a BT male set up residence on your property or in a nearby park? Even if that's not the case and you don't have the opportunity to watch a live mound, you may enjoy reading about these birds' intriguing pursuits of finding a mate and their unusual way of step-parenting. For many urban dwellers, BTs are a welcome bit of nature in the city and suburbs, and people appreciate the opportunity to watch their antics. The following will give you a better idea of what to look out for and when to best observe these ancient avians during the breeding season.

I am afraid that, as with most birdwatching, you will have to get up early to catch the bulk of BT mound courtships. We don't know exactly why females prefer to lay their eggs at this early time of day, but maybe it is because that is when the egg is usually ready to be laid or because there are less predators such as cold-blooded goannas or night-hunting owls around. Late risers may only be lucky if a female BT has not yet had a chance to lay her egg and visits a few hours after sunrise. This might be because she has been chased away from the mound at sunrise by other females — or even the male mound-owner. Let's find out why.

First thing in the morning, the male descends from his roosting tree, checks the temperature in his mound then adds or removes material to ensure his incubator is not too cold or hot. He needs to do so carefully, as mounds with stable incubation temperatures of 32.5°C to 34.5°C (90.5°F to 94.1°F) receive not only more eggs from different females but also larger eggs — and chicks from large eggs have a better chance of survival (see Chapter 14)[1]. As soon as he sees a girl nearby, he springs into action to attract her to his mound. Why exactly she chooses him we don't know, but she may be impressed by his 'prostrate' display, which he only shows in the presence of a girl. For this, he lowers his body to the ground, spreads his wings out to the side and frequently pecks at imaginary items on the ground without feeding. This display may be to pretend food is available or to indicate his mound is warm enough, and it certainly attracts the girls.

Females carry only one fertilised egg at a time that takes two to five days to develop until it is ready to be laid for incubation[2]. When they are on a mission to find the best nest, they may visit multiple males' mounds to dig a hole and check the temperature with their beak sensor. If it's not good enough, they move on to another male and mound. The females are highly selective when finding a place for their precious offspring to develop. As a result, they may fuss around on a mound for quite a while, digging a hole here and there and, even after all that testing, they may walk away without leaving their egg behind.

Such female procrastination is not to the male's liking. He wants her to lay her egg, not open up too many tunnels where the heat can escape, and he also wants her to leave so that other females can deposit their eggs. This may be one reason why the male is not all that welcoming to a female when she goes to test his mound. After luring her to his mound with his prostrate display, as she approaches he runs towards her and sometimes even pecks her. She spreads her wings to shield herself against assault while digging her tunnel in his mound. The male's aggression towards females on the mound is sometimes so high that we wonder about his motivation. Why does he not welcome the girls who want to lay an egg into his mound with open arms (wings)? David Wells at Macquarie

University tried to find an answer to precisely this question in his PhD studies, but even he had no foolproof explanation for it[3]. His most likely answer is that the male harasses the female so much because he wants to pressure her to mate with him (sexual coercion) in exchange for burying her egg in his mound[4].

Having to mate with the mound owner is something the female must accept, which leads us to the intriguing sex life of BTs. The male tries to mate with every BT girl approaching his mound and occasionally attempts to do so a little further away. The female has to pay this entry price to the mound. But the male can't really mate with a female when she is at term with a ready-to-be-laid egg, as she would eject his sperm when depositing the egg. So, he has to mate with females sometime after egg laying to ensure that the next egg she lays in any male's mound is his genetic offspring.

Through genetic studies, Sharon Birks found that on average 28% of the chicks from a mound had not been fathered by the male tending that particular mound. In one mound, this number even rose to 44%, meaning the females had visited many other mounds and males before laying their eggs in the mound where the genetic testing was done[5]. In summary, males are under a lot of pressure to mate with all the girls that visit their mound.

Egg-laying is hard work for the female as the egg is so large and must pass through a narrow cloaca to exit the body. She first digs a tunnel between 40 and 150 centimetres (16 to 59 inches) deep, depending on where the temperature is suitable. She then squats into the hole, spreads her wings to the sides and presses hard. The male watches attentively. As soon as the bright white egg is laid, it must be covered with mound material. The female occasionally does so, but the male often chases her away and takes over. Both males and females cover the egg with amazing tenderness, given that BT claws are sharp and could easily damage the egg during raking. With very small movements of the feet, she or he pads down some softer, decomposed soil around the egg. For a minute or two, they almost dance on the spot while quickly and carefully padding down the soil. David Booth named this the "stampy dance". Only once that is done does the male start raking with his usual powerful movements to heap more mound material on top and insulate the egg against outside temperature changes and rain during the next seven weeks as it develops into a chick.

All of this — luring the female, the temperature testing and egg laying, the mating and covering of the egg — may take up to ten minutes, and the male wants this to happen as quickly as possible. After all, the day is growing warmer, and if it is his mission to mate with as many females as possible he must ensure the mound is free for the next girl to come and lay her egg. On some mornings, you may observe several females on or near the mound and the male hectically trying to mate with them, observe them lay their egg, chase away those that are done or not serious, and possibly even having to deal with other males trying to interlope and be part of the fun. He's a busy BT indeed, and by mid-morning or lunchtime he is usually ready to have a rest in a nearby tree.

If we summarise this mating system scientifically, we call the BTs' behaviour both polyandrous (one female mating with many males) and polygynous (one male with many females). It explains why there is lots of chasing around the mound: the male chases away girls who are not serious about egg laying or who have already deposited their egg, and a female chases away other girls to gain access to her favourite mound. In the old times, when scientists described animal behaviour with an anthropocentric view, they would have called the BT mating system 'adultery'. Nowadays, we accept that animals behave in a way that maximises both their survival chances and their reproductive success. For BTs, this means the abandonment of monogamy and sexual exclusivity in favour of polygamy.

INTERVIEW

EMILY MANN
RESIDENT, NORTHERN BEACHES, SYDNEY

CAN YOU TELL US A LITTLE BIT ABOUT YOURSELF?

66 My name is Emily Mann and I live in the Northern Beaches, where my house backs onto Ku-ring-gai Chase National Park. 99

WHAT HAVE YOU LEARNED ABOUT LIVING WITH BRUSH-TURKEYS?

66 Growing up, there weren't many brush-turkeys in my area as the goannas kept the numbers down. Today, both the turkey and goanna populations are booming. The goannas have no predators and are clearly well-fed on turkeys, and the turkeys are either successfully breeding or more are moving into the National Park from distant suburbs.
I'm immensely fond of turkeys. They often join me while I do yard work and occasionally "help" by raking the leaves out of the roof gutters. They are so curious and fearless, which can be a good thing or a bad thing. Once a turkey turned up splashed all over in blue paint after knocking over someone's tins. The turkeys around my place chase each other constantly during breeding season and often have torn-off claws, head wounds, missing tails and bad limps. Their resilience amazes me.
They can be so goofy and awkward when they stand on the railings balancing their big bodies on top of their long legs. But then they will lower their heads and turn into battering rams to run and charge at other birds, asserting their dominance with a strangled gurgle. They crash-land directly above my bed at dawn and stomp across my tin roof like dinosaurs. They photo-bomb my pictures of other birds, steal food out of the hands of wallabies and generally have no boundaries with any human or animal. I adore them. They are instinctive survivors and lovable rogues.
99

The above observations are partly my own, but primarily based on Darryl Jones' ground-breaking work in the 1980s[6] and Sharon Birks' additional important observations later[7]. Darryl found that successful males attracted up to 12 different females to their mounds! At his study site in Queensland, some males even managed to build and look after <u>two</u> mounds simultaneously. They would run back and forth between both to ensure they would catch all the visiting females and mate with all of them.

While some males build two mounds, others are not successful in keeping just one. You may get all excited (or not) about seeing a male first build a mound in your garden, only to see this structure abandoned soon after. Darryl found that over half of all the mounds constructed were abandoned before receiving eggs, mainly because the males might have been younger and were chased away by more dominant males. Constructing big mounds is hard work, so some dominant males travel the easy way: they wait until a more subordinate male has done all the hard work then expel him and take over residency without much effort spent[8]. If you have a mound in your garden, you may be able to confirm Darryl's observation that mound ownership sometimes changes during the breeding season.

We have now explored the mound, when and how they are built as well as BTs' mating and egg-laying rituals. One question remains: how does building a mound out of tonnes of leaf litter affect the environment where the BT lives, especially the groundcover and plants growing in that area?

11

WHAT ARE THEY GOOD FOR?

THEIR ROLE IN THE NATURAL ENVIRONMENT

D<u>o BTs contribute to bushfire hazard reduction?</u>

Let's start this chapter with one of Australia's hottest topics: bushfires! This also means starting with the least proven impact of BTs on the natural environment, but one that nevertheless deserves consideration and further study. We know that BTs turn over a lot of leaf litter while scratching for food and building their mounds. This turning motion means the leaf litter decomposes more quickly and integrates into the soil. We also know leaf litter is fuel for bushfires and the more fuel load that is present, the more likely a fire will spread and increase in intensity.

The result of the BTs' continuous raking may well be a reduction in fuel load and an important helping hand in our fights against bushfires. The only study that has tried to quantify this was by Matthew Hall for his PhD thesis in 2022, when he found that leaf litter near BT mounds decomposed faster than further away, likely due to the BTs raking and digging behaviour[1].

While further studies on the decomposition of fuel load in the areas surrounding mounds are needed, we can also look at results for another Australian ground-raking bird: the lyrebird (*Menura novaehollandiae*). This large bird has a similar way of obtaining food by raking and digging in the litter on the forest floor.

Scientists at LaTrobe University in Victoria[2] have shown that lyrebirds reduce the fuel load in forests by about 25%. These researchers fenced off some areas where lyrebirds could not scratch then measured the amount of leaf litter in those areas compared to where the birds could forage as usual. They found that the fenced areas contained about 25% more leaf litter, obviously because the lyrebirds could not scratch there.

Brush-turkey and Superb Lyrebird

Size comparison of a brush-turkey and superb lyrebird. Both birds are so-called 'ecosystem engineers', meaning they shape the environment with their continuous raking.

Figure 17: A superb lyrebird next to a brush-turkey. Both birds are similar in size; the BT is just a little taller. Lyrebirds measure between 80-100 cm in length (including their elaborate tail), and BTs measure 60-70 cm but are heavier (1.2 kg in a BT male compared to 970 g in a male lyrebird).

Like the BT, lyrebirds continuously sift the forest floor, spread the leaf litter, bury leaf and other forest litter and, by doing so, speed up leaf decomposition and reduce the amount of fuel for bushfires. With all that impact on the ecosystem, lyrebirds have been named 'ecosystem engineers'. And to no surprise, BTs also own the same title, except we have less data available on how much they reduce the fuel load on the forest floor. It is safe to say, though, that they affect the amount of leaf litter available and are likely to play a positive role in decomposing bushfire fuel.

Increasing soil fertility

Apart from breaking down leaf litter, BTs also aerate the soil. The extra oxygen they add helps soil organisms to thrive. These small soil critters, in turn, break down the leaf litter even more, and as a result the areas where BTs scratch contain less leaf litter than other areas. In addition, looser soil means that nutrients and water can move through the soil more easily, and this also helps the plants and smaller organisms living and growing in that soil.

Distributing seeds

In the BTs' natural rainforest habitat, many native plants depend on some animals carrying their seeds to another place where they can grow with more sunlight than under the parent plant. BTs feed on various native seeds and fruit, especially in the rainforest. While digesting the fruit or seed consumed in one location, they move on elsewhere and spread the seeds with their droppings, playing an essential role in maintaining plant diversity.

Chasing snakes and other critters

Those worried about snakes in their backyard may want to encourage the presence of BTs. From various videos posted online, we know that BTs have little fear when chasing away snakes such as pythons. One video shows a fearless male attacking a diamond python (*Morelia spilota*) that slithers away from him[3], and I have also been sent a video of a BT chasing away a venomous red-bellied black snake (*Pseudechis porphyriacus*). Having a BT around definitely means less chance of having snakes nearby.

Due to their large size and varied diet, BTs feed on all kinds of other critters that some people don't want around the house. You may observe a BT breaking open rotten logs or other bits of wood to get to the termites inside – not many birds have claws that powerful to achieve this task! BTs eat termites (yes, those despised wood-destroyers), ants, centipedes and spiders. As a result, having a BT in your backyard means bug numbers are kept down.

Part of the natural food chain, important for biodiversity

While adult BTs may chase away a snake, they themselves are often eaten by another native reptile: the lace monitor (*Varanus varius*), a large lizard commonly called a goanna. I have been told of a goanna devouring an adult-sized BT, but it was unknown whether this individual was killed by the reptile or already dead beforehand, as goannas often feed on deceased animals (carrion).

We also know BTs are prey for the powerful owl (*Ninox strenua*), a large and rare species of native owl listed as a vulnerable due to fragmentation and loss of its habitat; wildlife managers are trying to encourage powerful owl populations back into our urban environments[4].

So far, we have talked about the role of the adults and how they are an essential part of our bush environment and natural food chain. What about the chicks — how do they survive all on their own?

INTERVIEW

JUDY HARRINGTON
BIRD LIFE AUSTRALIA

WHAT IS YOUR ROLE?

66 I am now retired but was one of the longest-serving ranger for the Sydney Olympic Park Authority. I am still an active member of BirdLife Australia and involved in many bird-related educational activities in Sydney. 99

SOME PEOPLE ARE ASKING: WHAT ARE BRUSH-TURKEYS GOOD FOR? WHAT IS YOUR RESPONSE TO THIS?

66 The first thing that comes to my mind is FIRE. I read about those studies on lyrebirds in the news, which showed that lyrebirds turn over so much leaf litter and help prevent fires by reducing the amount of it. Given that brush-turkeys have such a similar feeding behaviour, they surely must also play a similar role as 'fire wardens'!

When I see them feeding on all the fruit in the bush it is also pretty obvious that they help distribute the seeds of many native plants. These birds seem to be eating almost everything, and when they eat the fruit, they take the seed to another place and leave it there eventually, helping a new seedling to grow.

They also have an important role to play in nature's food chain. My son in Allambie Heights has seen a powerful owl that had caught a brush-turkey, and we really want to encourage those rare owls in the city.

In my many years as a guide for bird tours, I have noticed how important it is for city people to be able to observe birds and to engage with wildlife in the city. Brush-turkeys are unique and offer such a wonderful opportunity to observe a native bird that is easy to find and shows amazing behaviours. It's such a shame that if you put these birds in the wrong place, people regard them as weeds rather than the unique Australian bird they are. 99

INTERVIEW

JOHN KANOWSKI, AUSTRALIAN WILDLIFE CONSERVANCY

WHAT IS YOUR ROLE?

66 I am the Chief Science Officer at the Australian Wildlife Conservancy, a non-profit organisation dedicated to the effective conservation of all Australian wildlife. 99

WHY IS IT IMPORTANT TO LOOK AFTER MORE COMMON SPECIES LIKE THE BRUSH-TURKEY IF THERE ARE SO MANY ENDANGERED SPECIES THAT NEED OUR ATTENTION?

66 Brush-turkeys have been present most places I've lived – in rural Queensland and NSW, in the big smoke (Brisbane), and in the bush block where I am today. In all these places, they've been both endearing (fearlessly wandering across the yard) and annoying (self-appointed boss-chook and chief gardener). Australian forests (over east) wouldn't be the same without them.

Of course, brush-turkeys have declined from parts of their former range due to broadscale clearing and predation – their cute little chicks have to fend for themselves in a world of foxes and feral cats. In recent decades, however, they've re-established populations in the rich coastal ecosystems on which our cities sit.

So, how much should we care for the brush-turkey? Is it a big wheel in the ecological machinery of Australian nature? Or is it just a curiosity, a weird black bird (with a colourful head) smacked hard on the arse by the creator when she sent it on its way, flattening the tail feathers into a fan; a grunter not a singer; a robust raker of leaves, not a delicate weaver of nests?

Aldo Leopold, an influential thinker in the environmental movement, said "to keep every cog and wheel is the first precaution of intelligent tinkering." By this view, it doesn't matter what we humans think of the brush-turkey; we need to keep it anyway. 99

12

CHICKS COVERED IN SOIL

NOW, THAT'S UNIQUE FOR FEATHERED FRIENDS!

How BTs start their life after hatching contradicts almost everything we know about young birds. We have already discussed their unusual nest, but what about a hatchling that emerges from its egg on its own into soil instead of fresh air? How does it cope with being buried up to 1.5 metres (5 feet) deep and not receiving any help from its parents? How long does it spend in the soil, and how does it manage to survive and dig itself out?

Behaviour while buried in the mound

These were some of the questions I first tackled when I started studying BT chicks and to do so I had to observe the youngsters in the soil[1]. The most feasible method was to build a large transparent 'digging box' made of perspex. It was filled with leaf litter and soil from the core of a natural incubation mound, heated at a constant temperature of 34°C (93.2°F)and contained a sliding door. Once I found a chick that had just cracked its shell in my incubator, I immediately opened the door and placed this egg with its chick into the bottom of the box's soil column. With the help of some student volunteers, we then observed the chick in a dark room and only used a faint light to see what was happening inside the box.

Time-lapse cameras were unaffordable and tricky to use in the dark, so we relied on good old-fashioned real-time observation.

The first unusual thing we observed was how the chick hatched from the egg. Birds usually open the shell with their beak first, but the young BT cracked it open with its strong legs and pushed its back against the shell from the inside. With its head still stuck inside the egg, it kicked with its legs and trampled free a little cavity. This hollow was enough space to hatch into with the rest of the body and, apparently, it also provided enough air to start breathing. The whole process of hatching took several hours, and it was a very laborious exercise for the baby BT.

After the excitement of watching the chick hatch, my student volunteers couldn't wait to be the first people to ever observe a megapode chick in the soil. This enthusiasm slowly waned as time went by because nothing happened. We observers had trouble staying awake in the dark room while the chick rested. It didn't move much for the first ten hours, but it did change its appearance. At first it was still wet from all the fluid in the egg and its feathers were enclosed inside keratin sheaths. It also had some gel-like caps around its claws, which nobody had ever described before. Both these caps and the feather sheaths fell off as time went on. The chick dried and started looking much more like the cute fluffy feather-ball we knew from photos. It was also breathing heavily, filling its lungs with air. Occasionally, it would peck at items in its little cavity, such as small moving critters or twigs. A few hours later, it also started preening itself, using its beak to align its feathers and remove the feather sheaths.

After 16 hours of resting, the chick started digging its way up but only covered a few centimetres in 21 hours. It rested a lot between short bursts of digging in a little cavity that it always kept free around itself. My student volunteers were quite bored during this period, except when they observed the short digging bursts. The chick would push the top of its head and its back against the ceiling of the cavity, causing soil to fall downwards. It then sat on one of its legs while scratching along the side and ceiling of the cavity with the other foot. Finally, it would trample down the soil it had loosened underneath its feet.

Only after 37 hours — the students had already consumed a lot of chocolate bars to keep going — did the real action start. The chick was ready to face life outside, except more than 30 centimetres (12 inches) of soil was still above it. Using the same movements as before, it dug furiously and took only three hours to reach the top. It was breathing heavily and often completely covered in soil without a free cavity around it. Just before it fully emerged at the top, it stopped and rested again for five minutes. It seemed to listen carefully for any potential predators lurking in the big wide world before it suddenly dashed out and tried to run away. The box stopped it from doing so but, in the wild, the chick would probably have run straight into thickets to hide.

On average, it took the chicks 40 hours to reach the surface. Some managed to get there in 26 hours and others took 55 hours. How long they spent in the soil seemed to mainly be determined by how long they needed to rest after hatching and when first digging upwards.

Adaptations to being buried in the mound

My students and I still had a few 'why' questions to answer, and we did so by browsing through the literature or, where nothing was known, by finding the most obvious explanation. The latter was necessary to explain why the chicks hatch with gel caps around the claws. These seem to prevent the chicks from injuring themselves while they kick so heavily with their feet to emerge from the shell and form a cavity. Maybe they even serve the same purpose while the chick is still inside the egg, as the head rests right next to those big claws while the embryo develops.

An embryo inside the egg

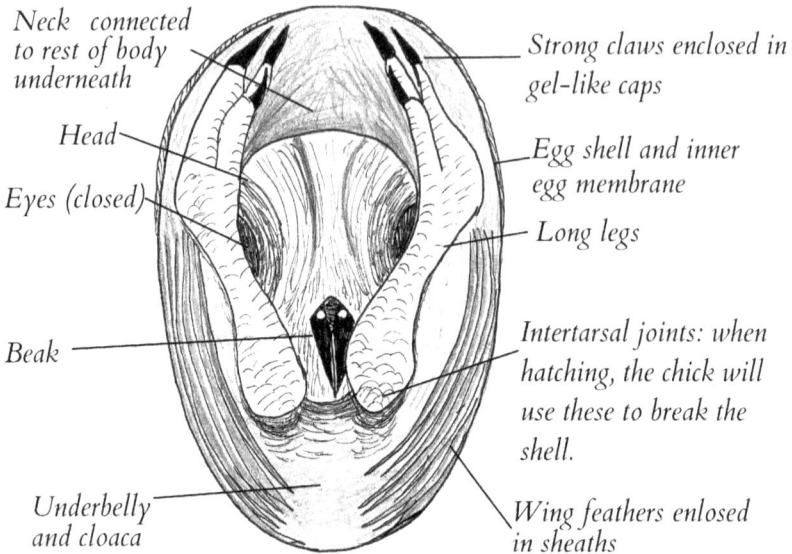

Neck connected to rest of body underneath

Strong claws enclosed in gel-like caps

Head

Eyes (closed)

Egg shell and inner egg membrane

Long legs

Beak

Intertarsal joints: when hatching, the chick will use these to break the shell.

Underbelly and cloaca

Wing feathers enlosed in sheaths

Figure 18. Position of a fully developed BT embryo inside the egg. Drawn from a photo of an embryo found dead in its egg inside an incubation mound shortly before hatching.

We also used our observations to explain why the feathers of BT chicks don't get damaged in the soil. Without clean, undamaged feathers a chick will not survive, as this coat will keep it warm, protected from rain and wind, and camouflaged so that predators are less likely to find it. It also needs the feathers to fly, escape predators and go up a tree to roost at night. BT chicks have three problems when it comes to their feathers. First, when they hatch their feathers are still wet from all the egg fluids and get muddy more easily. Second, the dirt around them is not conducive to clean feathers. Third, BT chicks, unlike other birds, hatch with their wing feathers already fully grown and developed, enabling them to fly as soon as they have reached the surface. The rest of their body is covered in fine hair-like down feathers that we recognise from domestic chicken chicks. They give the chick a fluffy appearance. While these fine feathers dry quickly if wet, the wing feathers (scientifically called 'primary flight feathers') don't. They are intricate structures made of fine barbs and can be more easily damaged.

Nature has equipped BT hatchlings with a unique way to protect their feathers: keratin sheaths. All birds have protective keratin sheaths (sleeves) around their feathers when they first grow them but these gradually fall off as the feather develops. In BT embryos, however, these sheaths stay on the feathers as the birds grow inside the egg, and they also stay on after hatching until the chick has dried and can preen them off with its beak. Once they are off, the chick is kept warm by the feathers and can regulate its body temperature well[2]. It also makes sense that the chicks hatch without proper tail feathers, only some down-like fluff at their rear end. This way, they don't have to worry about damaging tail feathers in the soil. Instead, they grow them about two weeks after reaching the surface (see Chapter 6).

While we observed some features that other bird hatchlings don't have, such as feathers fully enclosed in sheaths and plugs around the claws, one structure was absent in BT chicks. Baby chickens, for example, hatch with a little sharp lump on their beak called the 'egg tooth'. They use this for breaking out of the shell (pipping), and it is lost soon after hatching. BT and other megapode hatchlings don't have an egg tooth. Instead, they break the shell with their back and legs. But interestingly, their embryos, during the first few weeks of development, still have an egg tooth, only it disappears after that[3]. Scientists see this as evidence that over the course of evolution, the ancestors of BTs and other megapode did originally hatch the normal 'avian way' and that hatching the 'megapode way' developed later on as a secondary feature.

Chick digging itself out of the mound

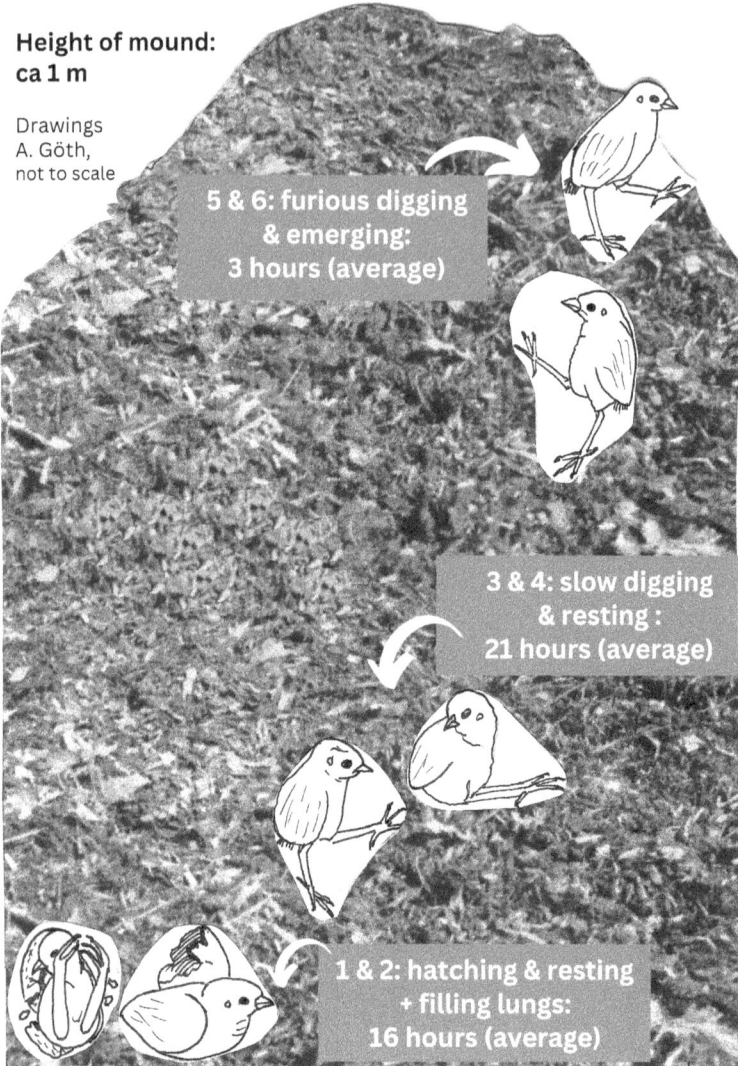

Height of mound: ca 1 m

Drawings A. Göth, not to scale

5 & 6: furious digging & emerging: 3 hours (average)

3 & 4: slow digging & resting : 21 hours (average)

1 & 2: hatching & resting + filling lungs: 16 hours (average)

Cross-section of a mound showing how one chick goes through different stages while digging its way out after hatching.

Figure 19. A chick is digging its way to the surface, showing the time it spends at different stages.

The heavy breathing we observed in the digging box was explained by reading the work of Roger Seymour[4]. He pointed out how BTs fill their lungs with air using an approach very different from other bird hatchlings. I must first explain how bird lungs work. They are not like the two expandable sponge-like structures we humans have. Instead, they are a more or less rigid lung that is ventilated by several other small cavities called air sacs. The reason for this sophistication is that flying requires a lot of energy, and the bird's unique lung system allows for a constant flow of fresh air through its lungs, ensuring it gets plenty of oxygen all the time. For BT chicks, this means they have to fill all those air sacs with air after hatching, as there was no air available inside the egg beforehand. Other birds have a little air bubble inside the egg from which the chicks start breathing even before they hatch, but BTs don't have that available either (see Chapter 14). So as soon as they have emerged into that little cavity, they need to suck whatever oxygen is available into their lungs and air sacs, and this may partly explain why they need to rest for so long after hatching and why they breathe laboriously.

Do they have enough air down there?

Where does the oxygen come from for breathing? And why don't they suffer from carbon dioxide poisoning? We know that when microbes decompose leaf litter they produce large amounts of carbon dioxide, a poisonous gas in high concentrations. Yet BT chicks buried in mounds manage to breathe well, and the chicks in our digging box did not seem distressed at all. There appears to be sufficient oxygen in all the small spaces between the leaf litter, sticks, twigs and soil incorporated into the mound. Somehow the oxygen moves (diffuses) into the small cavity where the chick rests and wriggles around. The movement of oxygen is also assisted by the adult birds when they rake and mix the mound material daily, preventing compaction and creating small cavities filled with air.

What do they eat while buried in the mound?

We also observed the chicks pecking at items in the digging box. Do they need to eat while buried in the soil? No, not really. Their pecking was more an exploration of their surroundings because their mother had equipped them with a decent packed lunch. While BT parents do not directly care for their young after hatching, there is one thing that mothers do when producing the egg: they add a considerable amount of yolk to it! The egg contains an enormous amount of yolk, much more than in most other bird eggs (see Chapter 14). At first, this yolk serves as food for the embryo during the seven-week-long incubation period, but the developing young does not use it all up. Just before hatching, it sucks up all the remaining yolk into its belly, and it is this packed lunch that it lives on during the first day or two after hatching when stuck in the soil. BT mums may not be so negligent after all!

The details for the chicks in the digging box above, including the times spent at the different stages, are averages for the 31 chicks we observed one by one in the digging box. I applauded my student volunteers for their perseverance and for staying awake, and we all walked away with deep admiration for a bird that hatches buried in soil and excavates its own path to the surface. We also knew that the books we had read were wrong. They all claimed that megapode chicks can fly as soon as they hatch. Flying is obviously not possible in the soil, and nobody knew that the chicks take THAT long to emerge from their underground nest. We forgive those authors for their lack of knowledge. Finding this out took patience, perseverance and many chocolate bars!

We now know how the chicks reach the surface, but how do they cope with being alone, without parents, after that?

13

WANT A SELF-SUFFICIENT KID?

CHECK OUT THE MOST PRECOCIAL BIRD

A BT chick sticks its head out of a mound. What is its first thought? Probably, "Go hide!" This may be followed by "What's there to eat?" But there are no parents available to hide it under their wings, nor to show it food or how to tell friends from foes. These youngsters need to manage on their own. For some scientists, including myself, they are fascinating birds to study as they can provide insights into how much learning is involved for individuals to recognise food, predators, roosts and members of their kind[1]. Citizen scientists, too, are fascinated by how a young bird can grow up without any parents, a topic that has featured on TV and in various magazines and newspaper articles[2]. More than once have other mothers commiserated with me about how much easier our lives would be if our offspring were that independent and far advanced (precocial) at birth!

Your first thought may be, "Well, there can't be much learning happening, given they have nobody to learn from", and you are quite correct. They don't usually hang around adults and live independently, so who teaches them what to do? They still need to learn certain things by trial and error or copying others occasionally. Let's look at their different behaviours: are they inbuilt (innate) when chicks hatch, or is there learning involved?

Recognising predators

To observe their response to predators, I moved two-day-old chicks into a large outdoor aviary and presented them with predators, including a live dog (Australian cattle dog) and a tabby cat. Both had been trained to walk through the aviary secured by a harness to a rope-and-pulley system. I also showed the chicks a rubber snake that I pulled along the floor and a silhouette of a bird of prey that 'flew' overhead. The youngsters crouched when they saw the bird of prey and they ran away quickly when the dog approached. The snake evoked the slightest response, as if the chicks knew they could escape quickly if they wanted to. However, their response to the cat was disconcerting: they crouched to the ground and only tried to run away when the cat was very close. The cat could have easily grabbed the chick if she had not been trained and secured with a harness. It appears that baby BTs do not have the appropriate inbuilt response to an introduced predator like a cat[3]!

Recognising each other

How do BT chicks recognise their own kind? How do they know what other BTs look like, given they have no mirror to view themselves in? In captivity, they form groups with other chicks from an early age and, even in the wild, a few chicks are occasionally seen together. The usual process by which birds learn to recognise conspecifics (members of their own species) is called imprinting. It describes how hatchlings form bonds to the first conspicuous object encountered, usually a parent. However, BT chicks do not imprint, as Sharon Wong at Griffith University discovered[4]. They must therefore respond innately to specific cues in conspecifics. Identifying these cues was the focus of a project I undertook with Chris Evans at Macquarie University. We pried into the minds of freshly hatched chicks with the help of a robot. In an outdoor aviary, we presented chicks with a choice between a static stuffed chick (made from the skin of a chick that had died naturally) and a robotic chick (made from the same skin and a little servo engine for remote-controlled cars). The 'robo-chick' either pecked at the ground as if feeding or scanned its surroundings by moving its head from side to side. Of these choices, the chicks most often approached the pecking robot, indicating that this movement was a cue for them to approach other chicks. It makes sense — these others may help find the yummiest food morsels around!

Apart from the pecking movement, another cue that helped chicks recognise their own kind was colour. Not just any colour, and not a colour that we humans can see, but UV colour. Birds can see UV well, whereas it is hidden from us. In a second experiment, we changed the colour of the pecking robot by mounting coloured filters above it. We found that chicks approached robo-chick significantly less if the UV, or short-wave colour, was removed from its feathers. By holding a spectrometer against them, we also realised that the chicks' beaks and legs reflect strongly under the UV. So chicks may also use UV reflection as another visual cue to recognise their own kind[5].

In addition, it would also make sense if they used the one-and-only call they utter to identify each other. To test this, Kate Barry and I presented younger (two-day-old) and slightly older (nine-day-old) chicks with the option of approaching either a loudspeaker or a pecking robot next to a speaker. We then played the typical calls of a BT chick from the speaker. The chicks did not approach either of these stimuli more than they did a control stimulus (e.g., a silent speaker or a pecking robot without a call), but they looked up and scanned their surroundings more often. So while they didn't seem motivated to approach another chick based on its calls, they may have recognised these as something worth looking out for[6]. Most of the time, baby BTs don't utter any calls and move quietly, hoping not to attract predators. This may also explain why they can't rely on finding other chicks based on their calls.

Recognising food

Recognising native predators and other BTs is clearly not difficult for chicks. But how about identifying food? In the wild, they feed on whatever invertebrates, seeds and fruit they can find. When Heather Proctor and I presented two-day-old chicks with various potential food and non-food items (such as pebbles), they soon learned to only peck at the edible food. Things that moved, such as earthworms, were their favourite, which is unsurprising given that their fast growth needs to be supported by a high-protein diet[7]. We also tested whether they preferred food that the pecking robot pecked at compared to food on its own, but there was no significant result this time. The chicks approached food on its own just as much as that presented with a pecking robot nearby[8]. So it appears that chicks don't need another BT to show them <u>what</u> to eat, but if presented with a robot that performs different behaviours, they pay attention to the pecking robot because it may show them <u>where</u> to find food.

Figure 20. A two-day-old chick ready to explore the world.

Chick survival

So now that we have identified that much of the chicks' behaviour is inbuilt (innate), we have to ask how this translates into their survival. Do most of them survive, given they essentially know what to do? The short answer is no. The long answer is partly provided by what we have already learnt about their response to cats — they duck down and wait for the cat to approach — which explains why so many of them are killed by our feline friends.

This was evident in a radio-tracking study I conducted with Uwe Vogel. We attached tiny radio transmitters to the chicks' backs with eyelash glue. A trial study had shown that these attachments did not affect the chicks' behaviour or ability to move, plus the transmitters fell off after about three weeks[9]. Altogether, we released 43 chicks with transmitters at a rainforest site and 33 in the bushland near a rural farm. The rainforest (Mary Cairncross Reserve in Queensland) was their natural habitat but contained relatively little groundcover (apart from the nasty, spiky lawyer vines that ripped at our skin and clothes). The nearby farm site had large patches of introduced weeds that formed thickets, such as lantana and blackberry. These thickets provided the best cover for the chicks, and they explained why, at the farm site, at least 12% of the chicks survived the first three weeks whereas, unfortunately, none made it to that age in the rainforest. Chick survival was higher in areas where these babies could hide away from cats and other predators.

What were the other predators? While domestic or feral cats killed most chicks (51% at the rainforest site and 29% at the farm site), some were taken by birds of prey (29% at the rainforest site and 10% at the farm site) and the remainder succumbed to dingoes, foxes, disease or unknown causes. How did we know? When we inspected the feathers or carcasses of chicks with the transmitter nearby, we knew that cats chew off their feathers, birds of prey pluck them out so the feathers are not damaged at the base, and dingoes and foxes bite off whole wings, leaving clumps of feathers largely intact[10]. We could not tell exactly which birds of prey had killed our chicks, but we knew from anectodal reports that BT chicks do get eaten by large birds such as kites, hawks and owls, as well as kookaburras and Torresian crows.

The high rate of chick mortality has to be seen in relation to the large number of eggs these birds lay. In the animal kingdom, we call animals that produce lots of young but don't look after them *rapid strategists* or *R-Strategists*, while those that have fewer offspring but try to make sure they all survive are called *K-Strategists*. BTs are more of an *R-Strategist*, whereas most other birds can be called *K-Strategists*. We thus expect chick mortality to be quite high, though it is definitely unusually higher in areas where feral cats prey on chicks. With females laying up to 20 eggs per season, far more than brood-nesting birds, the world would be chock-a-block full of BTs if all the chicks survived.

Chick movements

At least some hatchlings made it through to age three weeks before the transmitter fell off. What did they do and where did they go? Radio tracking required us to physically follow the chick as often as possible to avoid losing the signal. We appreciated those chicks that stayed in the same thickets for many days, moving only five metres away from the release site. Here, they happily fed on whatever bugs and berries they could find, took sand baths to keep clean and rested camouflaged on the ground or in a higher bush at night. Other youngsters kept us on our toes as they dashed across open areas to reach thickets further away, some travelling up to 800 metres (875 yards) per day. The signal of their tiny transmitters could only be picked up from a few hundred metres away, so more than once we thought that we had lost these agile little ones only to find them again when we searched the area by car. Looking at the results for all the chicks we released, they travelled 100 to 200 metres (109 to 218 yards) on average during the first five days after hatching[11].

I still remember when we seemed to have lost the signal from one of our chicks. We searched far and wide, walking up and down our rainforest site, until we finally picked up a faint signal from the row of houses that bordered the site. Had the chick become one of those urban colonisers that move out of their natural habitat? We narrowed the signal down to one dwelling, and it clearly came from inside the house. It was time to knock on the door and ask the owner whether she had a cat. "Yes," she said, "I do, but it never kills any birds." We asked her to check, and there was the cat chewing on our chick on the expensive Persian rug in the living room, blood stains all over the carpet and the transmitter discarded nearby. That owner had no idea what her cat was up to when outside!

Where are the chicks?

Many people say that while adult BTs are a common sight, they never see any chicks. There are two reasons for this. First, as shown above, the youngsters like to hide under cover, where they can increase their chance of survival. Second, they grow up quickly and don't look like cute fluffy chicks for more than a few weeks (see Chapter 6). In some suburban settings, such as Taronga Zoo in Sydney, the chicks learn to leave their hiding spots early on, especially when food is on offer. Members of the public who spot baby BTs in their yards often call up wildlife rescue lines to report an abandoned chick that needs help (see Chapter 18). Hopefully, readers now know better: the apparent helpless fluffball is, in fact, one of the most self-sufficient kids in the bird world!

But what about before they hatch? How do BT chicks cope when they are still in the egg? They may only be embryos, but BT embryos face many challenges that other birds don't have to worry about.

14

Egg-citing Stuff

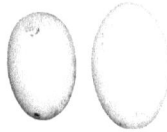

Now that's a special bird egg!

While BT mothers have no say in raising their babies, one thing they can influence is the egg. Think of the egg as the womb in which the mother provides sufficient space, food and even hormones for the chicks to develop properly, and where the cells of the embryos have enough oxygen to grow into chicks. Except, with BTs, that womb is outside and dumped into a big mound of dirt full of bacteria. So before leaving the egg to be incubated in a pile of rotting vegetation, BT mums must pre-pack the egg-womb with everything the embryo needs. Let's explore what this encompasses.

Egg colour, shape and shell structure

Pure white is the best colour to describe the shell of a freshly laid BT egg. Once in the soil, it doesn't stay speckless clean for long, as the dirt and leaf litter surrounding it will soon stain the shell a brownish colour. This means that older eggs can be distinguished from the unspoiled white of freshly laid ones. Eggs are almost symmetrically oval in shape, with a slightly blunter end on one side and a pointier shape on the other. The pointed end makes it easier for the female to push the large egg out of her constricted cloaca during laying and is thus usually facing downwards in the mound (unless the egg is moved out of position later on).

While none of this distinguishes a BT egg much from an ordinary chicken egg, there are two huge differences that you can only see when inspecting the structure of the eggshell with a microscope. David Booth from the University of Queensland found that the shell is much thinner than in other birds, about 31% thinner than a similar-sized egg in other chicken-like birds[1]. Not only that, but the shell becomes even thinner during the incubation process because the growing embryo incorporates some of the shell's calcium into its bones, and the shell's inside is gradually dissolved[2].

Not only is the shell extra thin, but it also contains special types of trumpet-shaped pores (microscopic holes in the shell). When bird embryos breathe, they take in oxygen while producing the by-products of water and carbon dioxide, a gas that is poisonous to the embryo if it is not removed from the egg. Thus, for them to develop, the excess water and carbon dioxide must move out of the egg while oxygen must enter for the embryo to continue to breathe. Unlike other birds, BT eggs are incubated in an environment with very high humidity and extreme gas conditions. Under such circumstances, the essential exchange of gases in and out of the egg is only possible if the eggshell is extremely thin and contains special pores that let these gases diffuse in and out of the egg more easily than the shells of other bird eggs[3].

Liliana D'Alba and her colleagues discovered the second surprising difference between BT and chicken eggs[4]. They found that the BT eggshell is covered in tiny nanoparticles of calcium phosphate that prevent nasty, sickening soil bacteria from sticking to the shell and getting through the thin shell layer to the chick. D'Alba demonstrated that more than three times the amount of bacteria adhered to the shell of chicken eggs than to BT eggs. This is an amazing adaptation of eggs incubated in the soil, where the bacterial load is much higher than under a brooding hen. It was such a surprising discovery that it was even featured as a spotlight in *Nature*, one of the world's most renowned scientific journals for the natural sciences[5].

Incubation period and temperature

"Don't count your chicks before they hatch!" If you are referring to chickens, you shouldn't make any plans until 21 days have passed. But be prepared to have extra patience if you're referring to BTs — their incubation time is more than twice as long as chickens! In fact, their incubation period is longer than most other birds.

On average, BT chicks hatch 49 days after the egg has been laid. This incubation period can be a few days shorter or longer, depending on the temperature in the mound. Lower temperatures mean a longer incubation time and higher ones cause the chicks to hatch earlier. Not much earlier, though, as prolonged incubation is needed to produce a highly developed chick at hatching[6].

Let's explore at what temperatures the eggs are incubated. The males are not always able to keep the temperature 100% stable, as the incubation mound is exposed to sweltering days as well as the cooling effects of rainfall and colder weather. Despite this, several studies in different parts of Australia's East Coast found that the average incubation temperature was between 33°C and 34°C (91.4°F and 93.2°F)[7]. For my studies, I measured the temperature next to 339 BT eggs[8]. Even though most eggs were found to be incubating between 32.5°C and 34.5°C (89.6°F to 94.1°F), the mound temperatures could range between 27°C and 37°C (80.6°F to 98.6°F). A recent study by Yvonne Eibing from the University of Queensland also found most eggs deposited at these same temperatures of 32.5°C-34.5°C but reported a much wider range of incubation temperatures next to eggs than previously known. She observed embryos in the mounds thriving despite prolonged exposure to sub-optimal temperatures over the range of 25°C to 40°C (77°F to 104°F)[9]. Not many other birds can cope with such challenging variations in incubation temperatures.

Egg size and egg numbers (clutch size)

BT eggs are huge. On average they weigh around 180 grams (6.4 ounces), about 10% of the female's bodyweight, and the female lays a fertilised egg every two to five days. Females need to find a lot of protein-rich food to build such a big egg. Producing one at a time is certainly enough for them, and they do so over the whole breeding period of approximately five months. Based on the observed interval of two to five days between eggs, they produce around 18 to 24 eggs per breeding season; more impressively, this represents almost three times their own bodyweight in eggs every season[10].

Other birds lay a particular number of eggs (called a clutch) and start incubating all the eggs in their nest at once. This way, they ensure that all the embryos develop at the same rate and hatch simultaneously. BTs, by contrast, lay multiple eggs at different times for incubation under the soil. This means the eggs found in a mound are all at various stages of development unless two females lay their eggs on the same day, and it also means that chicks hatch continuously throughout the breeding season, each on its own timeline. I once found 50 eggs in the same mound later in the breeding season, all at different stages, from freshy laid to almost ready to hatch!

Variation in egg size

Just like humans, not all BT babies are the same size. With females having to find so much protein-rich food to produce their large eggs, it is no surprise that not all girls can create the biggest possible egg each time. For my studies of chicks, I uncovered 263 eggs (and later returned all chicks to the where they had come from) and I found that the largest egg weighed about twice as much as the smallest one[11]. Only about 3% of other birds have such a significant difference in egg size!

Apart from not finding enough protein-rich food, females may also lay smaller eggs simply because they are smaller or not 100% fit and healthy. Why is this worth mentioning? Because egg size considerably affects the offspring. Chicks from small eggs are smaller, dig themselves out of the mound more slowly, grow at a slower pace and are least likely to survive[12]. Hence females are under pressure to produce the largest eggs possible. We also have some indication that those girls who do manage to lay big eggs are also the ones who have access to the best incubation mounds, as I found that the mounds with the best temperatures (those closest to 33° C) contained the largest eggs[13].

Figure 21. Photo of two BT eggs that varied considerably in size. Both of these eggs were taken from the same mound. Photo Ann Göth.

Proportion of yolk

At the beginning of this chapter, I mentioned that females can best assist their kids via the egg, as they don't have much say after the chick has hatched. The egg yolk is the female's one provision that is most important to their chick after it hatches.

What makes a lovely yellow omelette for us has a different importance for bird embryos. The yolk is the food needed while they develop into a chick. It provides them with energy and nutrients such as protein, fat, vitamins and minerals. Considering that a BT embryo develops for much longer than most birds — seven weeks on average — it needs more yolk than a young chicken that hatches at three weeks. Accordingly, BT eggs contain much more yolk than most bird eggs: around 50% of the egg is yolk compared to 37% in chicken eggs[14].

The yolk not only supports BT embryos during incubation, it also serves as a snack for the first few days after hatching. Remember, the chicks take almost two days to reach the surface (see Chapter 12), so having Mum's packed lunch available during that time is handy!

Egg comparison

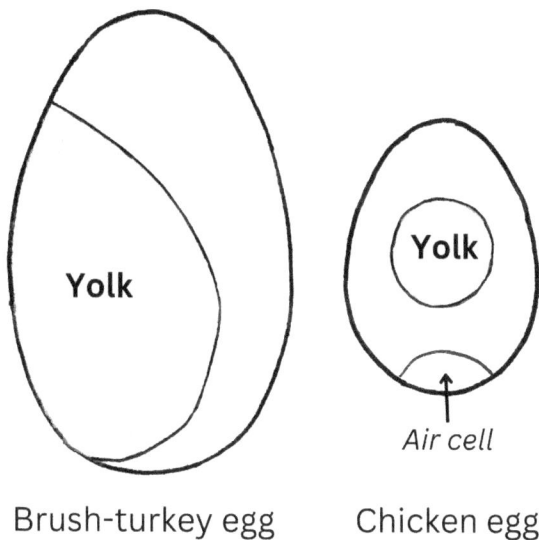

Yolk

Yolk

Air cell

Brush-turkey egg   Chicken egg

Figure 22. Brush-turkey egg and chicken egg drawn to scale, allowing for a size and yolk content comparison. Also, note that chicken eggs contain an air cell for embryos to start breathing before they hatch, but brush-turkey eggs don't. (Brush-turkey egg drawn from the photo of an infertile egg that had been frozen and then peeled, chicken egg adapted from https://www.chickens.allotment-garden.org/eggs/structu re-egg/).

Egg hormones

In birds, the mother's hormones in the egg yolk affect the youngster's behaviour, looks, internal functions (physiology) and chances of survival. For example, more testosterone in the egg of a regular bird, regardless of its gender, means the chick hatches earlier, has bigger muscles and wins more often when it competes with its siblings for food from their parents. BT chicks don't have to worry about sibling competition, but they do have to dig themselves out of a large mound. Could the mother's hormones in the yolk influence how well they can do this?

A team of researchers from Macquarie University in Sydney and the University of Groningen in the Netherlands assisted me in studying yolk hormones[15]. They taught me how to carefully take a sample of yolk without damaging the egg or affecting the embryo inside. It can be done by inserting a thin sterile needle through the shell into the yolk then closing the hole in the shell with surgical tape. The chicks from the eggs I sampled developed as usual, and I could send the yolk sample to the Netherlands for analysis.

Two of the hormones the mother deposits in the egg are testosterone and androstenedione. (Even though you may think testosterone is only a male thing, females also produce it). We found that eggs laid in bigger mounds and at greater depth contained more of these two hormones than those eggs from smaller mounds and laid at less depth. With this study, the first one for megapodes, we only scratched the surface and many questions remain. Do mothers help the chicks that are buried deeper in the mounds by providing them with more testosterone and thus enabling them to dig themselves out more easily? We don't know, but it is an interesting thought.

Australia's Indigenous people also had, and still have, an interest in BT eggs. In the next chapter, let's look at the role of BTs and their eggs in the culture of Aboriginal people.

INTERVIEW

DAVID BOOTH
SCIENTIST, UNIVERSITY OF QUEENSLAND

WHAT IS YOUR ROLE?

66 For almost 32 years, I have been a scientist and senior lecturer at the Biology Department of the University of Queensland in Brisbane. I have studied the special adaptations of brush-turkey eggs and their unique incubation strategy. 99

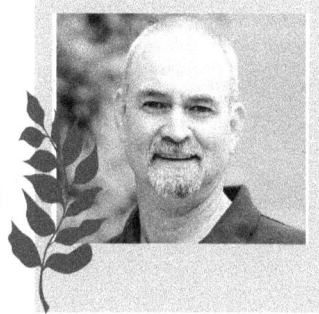

WHAT ATTRACTED YOU TO STUDYING BRUSH-TURKEYS?

66Brush-turkeys are fascinating because of their underground nesting habit. Burying eggs in incubator mounds means an individual female can lay many eggs (up to 20 eggs) over the long breeding season, rather than laying all her eggs at once. It also allows several females to lay their eggs in the same mound.

Of great interest to me are the adaptations of eggs to underground incubation. The eggs are about three times larger than a domestic chicken. They have a really big yolk, which supplies the extra energy needed to dig out from the mound after hatching.
The large yolk is also a food source during the first few days post-hatching as their parents don't feed them or show them where food is.

The eggshell is thin because they are not banged against other eggs in the nest or sat on by adults during incubation. The eggshell is also more porous than regular bird eggs, which makes it easier for the developing embryo to obtain oxygen and get rid of carbon dioxide across the eggshell, an adaptation to the relative low oxygen and high carbon dioxide levels within the mound compared to the open nest of other incubating birds.99

Section 3
The Power
of Knowledge
CHAPTERS 15–16

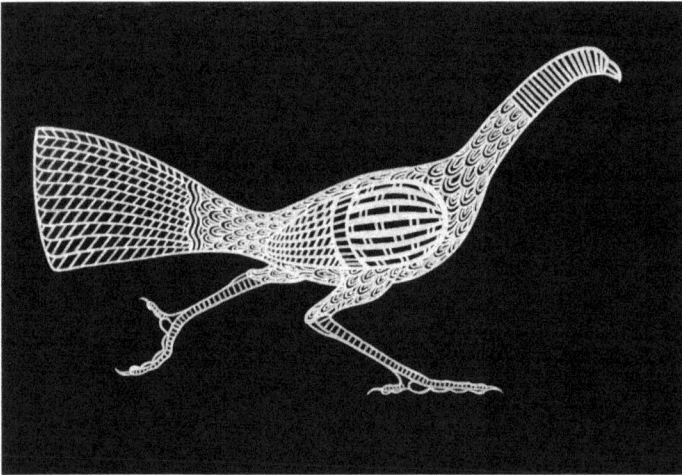

Aboriginal artwork of brush-turkey by Dr. Shane
Smithers, reproduced with his permission.

15

ABORIGINAL KNOWLEDGE

BRUSH-TURKEY LORE FROM THE ANCESTORS

B Ts hold a significant place in Aboriginal lore and stories, and are an important totem animal for many Indigenous clans. In addition, Aboriginal and Torres Strait Islander people have always welcomed BT eggs and meat as bush food.

Darug man Dr Shane Smithers is an artist and philosopher of the Burraberongal clan of Western Sydney[1]. Shane has given me cultural permission to share his ancestors' lore about these birds.

The BT, with its habit of hatching from a mound of earth, symbolises life emerging from the Earth and as such has an important ceremonial role as a symbol of creation for many Aboriginal people. The story of the BT is one of the oldest Aboriginal stories from the deep Dreaming, ancient yet relevant in every age.

Aboriginal people see the mound from which chicks hatch as representative of the Earth that gives birth to all life. The ground surrounding a BT mound is usually raked entirely clear of leaf litter and other materials by the male. The layout of many Aboriginal ceremonial grounds is inspired by the BT mound: an inner circle (representing the mound) surrounded by a ring of stones or other material (representing the bare ground). Aboriginal people imagine this is how a BT in the stars would view the ceremonial ground from above – like his own mound surrounded by barren land. A long time ago, the inner circle was replaced by an actual mound that was created as a burial mound. It was surrounded by an area in which the ground was swept clean and bordered by an outer circle made of soil. A picture of such a burial mound, which closely resembles a BT mound and the surrounding area, is depicted in a book about Aboriginal Carved Trees[2].

The star constellation we know as Scorpius is known as the Broken-Neck Turkey constellation by Shane and his ancestors. It has important meaning for the timing of men's gatherings, when older men take the younger men to special ceremonies. Aboriginal people carefully observe the Broken-Neck Turkey constellation in relation to the BT's breeding cycle to determine this timing. For Shane and his ancestors, the constellation also holds a prophecy about the destruction and eventual rebuilding of Aboriginal lore. This lore shared by Shane corresponds with the same story from the Ngemba people in New South Wales, written down in a publication about Aboriginal astronomy[3]. The following is a direct quote from an Aboriginal advisor (named P7) in this paper: "*Scorpius is the 'broken-neck' brush turkey. The tail of Scorpius is the head (twisted back); the head of Scorpius is the fan tail. The turkey is an important part of Law and ceremony. The brush turkey scratches the ground to make its nesting mound around June, which is when they have the Bora ceremony.*" The advisor, P7, said that the turkey has been engraved in rock in places, and there is an engraving of the turkey with a clever man pointing to it in the sky, telling a story which is a prophecy. P7 also said: "*Meteors come from Scorpius, and the turkey's neck was broken by a meteor.*"

In traditional Aboriginal culture, moral teachings, knowledge, wisdom, philosophy and spirituality are transmitted through stories, dance, music, paintings and rock art. Their oral tradition is not limited to speech or storytelling but includes visual language and dance movements. BTs are one of the iconic figures of traditional lore, together with other figures such as the eagle, crow, kangaroo and emu. These animals represent the entirety of lore given to all things, including places, the sky and the Earth. They also represent the means through which lore is transmitted.

Shane also kindly gave me cultural permission to reproduce one of his artworks of a BT that he painted in 2017.

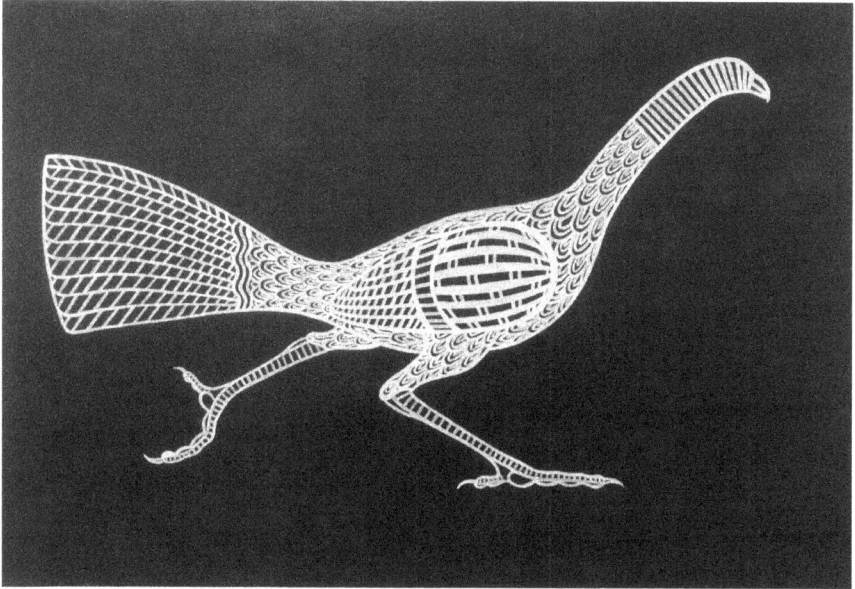

Figure 23. Shane Smithers 2017, Adult Brush Turkey (Lore of Country, Connection of Sky & Earth). Reproduced with permission from the artist.

In addition to the painting, Shane created a stone carving of a BT as a replica of an original carving that is much larger (approximately 1.5 metres, or 4.9 feet, wide). The original is found in Wollombi, an area near Mt Yengo in New South Wales, at an important meeting place for several Aboriginal clans that contains many rock carvings. The carving is clearly of a BT hatchling, as it does not yet have a tail.

Figure 24. Rock carving by Shane Smithers, resembling a larger carving (petroglyph) at Wollombi. It shows a hatchling, illustrating the miracle of life emerging directly from the Earth. This carving also relates to the Broken-Neck Turkey constellation. Reproduced with permission from the artist Shane Smithers.

Other Aboriginal artists such as Clifford Madsen have painted BTs in the well-known dot-painting style[4]. Be careful, though, when viewing Aboriginal artwork featuring BTs. Some of this art does not depict BTs but rather the Australian bustard (*Ardeotis australis*), a bird also called a 'bush-turkey' by some. It, too, is a bird important to Aboriginal people.

Aboriginal names for BTs

Different Aboriginal clans have various names for BTs. The most common names, *wagun* or *wirrilaa*, are used in several Aboriginal languages, including Yuwaalaraay, Gamilaraay, Gamilaroi and Kamilaroi[5]. The Kombemerris clan called them *woggan* after the sound they make[6]. The name *wee-lah* is given in a publication from 1861[7], and a 1919 book about Australian wildlife mentions the Aboriginal name *gweela* for BTs[8]. Up north, Torres Strait Islanders call them *surrk au,* and they make a lucky charm in the shape of BT legs, believing that it will bring them luck when they hunt, fish or search for wild yam[9].

The effect of incubation temperature

The most astonishing information I obtained from an Aboriginal elder about BTs was in the Atherton Tablelands in Queensland. Ngadjon-Jii man Warren Canendo informed me that his ancestors knew about the effect of the incubation temperature on the sex ratios of BT hatchlings. They observed more female BTs in the area following a warm breeding season, and more males, with their big wattles, following a cooler breeding season. This was news to any non-Aboriginal bird researcher or enthusiast. There was no account anywhere in the bird world of the incubation temperature affecting the sex ratios of the young. It was only known to occur in reptiles that also build a mound for incubation, such as in crocodiles.

Could this also be possible in a bird? Some pondering made me realise the similarities between BTs and crocodiles. They both build and bury their eggs in mounds, even though crocodile mounds are not as carefully constructed as BT mounds. Plus, BT eggs tolerate a wide range of incubation temperatures (see Chapter 14), just like in crocodiles. Given this similarity, David Booth and I researched this further. In the artificial incubators, we chose to incubate some freshly laid eggs at the cooler incubation temperature of 31°C (87.8°F), some at the average temperature of 34°C (93.2°F), and the remainder at the warmer temperature of 36°C (96.8°F). Indeed, more males hatched at the cool temperature and more females at the warm temperature. In contrast, the sex ratio was almost exactly half males and half females at the average temperature[10]. After this first experiment, I also measured the temperatures in mounds found next to some far-developed eggs then took those eggs to the lab to be incubated at those same temperatures to determine the sex of the chicks that hatched from them. I found the same pattern: the sex ratios of these chicks significantly correlated with the mean incubation temperatures in their original mounds[11].

Researchers who are making discoveries that contravene widely accepted knowledge are often a bit anxious about publishing their findings. David and I were nervous when we claimed that the incubation temperature could affect sex ratios in a bird. Additionally, we had based our research on the traditional beliefs of Aboriginal people passed down through generations. But we braved any potential criticism and were right to trust ancient cultural knowledge, as our results caused a bit of a stir in the scientific community[12].

Luckily, Yvonne Eiby, Jessica Wilmer and David Booth later confirmed these results in another study and, in even more exciting news, they identified <u>how</u> this happened. Through DNA analysis of embryos, they confirmed that at warm temperatures more male embryos died in the egg and when it was cooler more female embryos did not make it to hatching. The underlying mechanism was described as 'sex-biased embryo mortality'[13]. We don't know the purpose of this phenomenon, but it raises questions about the future impact of climate change: will we have more female BTs and not enough males if our climate keeps warming?

BTs as weather prophets

Another cultural insight is that Aboriginal people use BTs as prophets of upcoming rain by watching the males on their mounds. They say the males will start heaping more material on top of the mound to insulate it <u>before</u> it rains. We don't exactly know how a BT may predict the rain, but it is obvious why such precautions make sense. Some moisture in the mound is necessary for heat production, but if a mound becomes too waterlogged the eggs can't develop, or flood waters may even sweep away parts or all of a mound. The rain will reduce mound temperatures below the levels needed for incubation and stop chicks from developing to the hatching stage.

Aboriginal people also know other details about BT ecology and behaviour that nobody has studied yet. For example, the same Aboriginal elder who told me about the effect of incubation temperatures knew that, on the Atherton Tablelands, both BTs and the orange-footed scrubfowl, a megapode relative, sometimes lay their eggs in the same mound. How does this work – which species is building and defending the mound? What is the advantage of such a strategy, or the disadvantage for each of the two species?

The legend of Wollumbin

Aboriginal people have incorporated BTs into many of their legends and Dreamtime stories. One is how Wollumbin was created, a 1,157-metre (3,796-foot) peak at Mt Warning in New South Wales. The following is quoted directly from the Planet Corroboree website[14]: "*Once, a long time ago, turkeys could fly greater distances than they can today. Well, one turkey flew from Mt Brown. He had joined a gathering of other birds talking, when a giant bird approached. All the birds, the turkey too, were so frightened they flew away quickly. The turkey flew all the way to Wollumbin, stopping on the top of the great mountain to catch his breath, but as he rested, he was wounded in the head by a spear from a warrior. Because of his head wound, the turkey's flying ability was impaired. That is the reason why turkeys today can fly only short distances before they must rest; and if you look at Wollumbin you can see the mountain top has a small bend in it where the spear hit the turkey. The mountain tip is the point of the warrior's spear.*"

I am certain Aboriginal people have many more stories and insights into BT ecology and behaviour that we haven't heard yet. If any of my readers are aware of any of these, please let me know! In the next chapter, let's explore some other studies on brush-turkeys.

16

Other Brush-Turkey Studies

Why we have more insights to 'talk turkey' today

While Aboriginal and Torres Strait Islander people have accumulated stories and observations of BTs for thousands of years, modern scientists have done so only recently. Early English explorers travelling to the 'new' southern continent provided some general descriptions of BTs. One was John Gould, who published his famous book *Birds of Australia* in the early 19th century and included the first hand-coloured lithographic illustration of a BT, then known as the 'wattled tallegalla'[1]. Gould included the following poetic description of how BTs hatch (p. 98): *"As a moth emerges from a chrysalis, dries its wings, and flies away, so the youthful Talegallus, when it leaves the egg, is sufficiently perfect to be able to act independently and produce its food".*

Other scientists then called the BT the 'New Holland vulture' because of its bare head, and their scientific interest was to identify its closest relatives amongst the birds. While they couldn't agree on the same common name, English zoologist and taxonomist J.E. Grey officially set the BT's Latin name in 1831. He named it *Alectura lathami* after John Latham, a leading English ornithologist (bird specialist) who had just published the book *A General History of Birds*[2]. Interestingly, Latham had misidentified the BT as a type of vulture based on some drawings he had seen. He only admitted to this mistake in the ninth edition of his book, when it had become apparent that these birds were utterly unrelated to vultures[3]. For those eager to know why J.E. Grey chose the word *Alectura* for the Latin name, it means 'cock tail' and is derived from the Greek words *alektoris* ('cock' or 'chicken') and *oupá* ('tail')[4].

Overseas, there continued to be interest in the BTs' unusual breeding behaviour, such as the below 1887 illustration from a *Scientific American* publication showing a male building his mound. In 1908, some poor BTs and a few emus were shipped off to England, where bird specialists regarded them as a unique curiosity[5]. Despite the London climate, these tough BT males built an incubation mound in the London Zoo, leading to a 1933 description of their unusual breeding strategy in the renowned scientific journal *Nature*[6].

THE BRUSH TURKEY (TALEGALLUS LATHAMI)

Figure 25: Brush-turkey building his mound. From Mirtzel, G. (1887). *Scientific American*, 57(11): 162. No known copyright restrictions.

Meanwhile, in Australia, where non-Aboriginal people were struggling to settle and survive, the focus was on hunting and eating BTs, culminating during the Great Depression in the 1930s when other food was scarce (see Chapter 5). Publications focused on the best recipes for cooking them, with the exception of author Edward Sorenson who wrote about their behaviour in the 1919 book

Friends and Foes in the Australian Bush[7]. It contains an amusing story about 'Gweela, The Brush-turkey' and, given the time it was written, contains some surprisingly correct observations (as well as the incorrect assumptions that BTs live in pairs and dig themselves out as soon as they hatch). Here is a taste of Sorenson's rather poetic writing style: *"Unlike most birds, he did not come out of his shell under the sheltering wings of a mother. Such a creature he never knew. Nor did he chip his shell and call for help like other chicks when it was time to get out. He just gave a heave of his strong little body, and the shell, which had become brittle, burst into fragments, leaving him erect on his feet. He stood in inky darkness, with a great weight pressing all around him."*

In 1944, Alfred Russel included general descriptions of BT breeding behaviour in his book *Bush Ways*[8]. Russel's description of the chicks' behaviour reads: *"As for schooling, the chick neither knows nor needs parental care; schooling itself it instinctively adopts, as it grows to maturity, the life habits of its kind. To few animal folk has that power been given."*

In Australia, a short publication by David Fleay from 1937 described the nesting habits of the BT[9], while the first detailed scientific study of BTs did not occur in Australia but in two German zoos in the 1960s[10].

However, more thorough studies did not occur until 1983, when Professor Darryl Jones appeared on the scene. He was the first scientist to dive deeper into the hows and whys of BT behaviour. On Tamborine Mountain in Queensland, Darryl battled mosquitos and leeches to observe the BTs' peculiar behaviour on the incubation mounds, especially their social relationships. In the following years, he published his results in several scientific journals (see Chapter 10) and became known as a 'brush-turkey guru' or 'brush-turkey whisperer'.

From the early 1990s, Darryl studied a 'brush-turkey war' unfolding in Brisbane, where he observed a mass migration of BTs into suburbia that seemed like a mirage to him and many others – just as it is happening in Sydney today (see

Introduction). Journalists sought his expert advice on why the BTs were moving into the city and suburbs, how they survived and what people could do to get rid of them. Not even Darryl can recall how often he has 'talked turkey' on radio, TV and to newspaper journalists. To this day, he is still the go-to person for journalists seeking to interview a BT whisperer. If you want an entertaining and fascinating read of Darryl's BT antics, I highly recommend Chapter 4 of his recent book, *Curlews on Vulture Street*[11].

Darryl Jones did not just focus on his own BT research. He cooperated with researchers from other universities worldwide, sought funding for BT research and became an active member of organisations that promote the study and conservation of this bird and its megapode relatives. Darryl co-founded the Megapode Specialist Group, affiliated with the International Union for the Conservation of Nature, and co-authored the first book on megapodes, a monograph titled *The Megapodes*[12].

Darryl's tireless support and enthusiastic encouragement meant that I could come to Australia to conduct my PhD study on BT chicks, and I am very grateful for this. After several years of working together, we jointly published the book *Mound-builders* about the BT, malleefowl and orange-footed scrubfowl[13]. Darryl supervised other students who focused on different aspects of BT behaviour and ecology for their theses, including Susan Everding, Sharon Wong and Sharon Birks. His busy career at Griffith University in Brisbane also involved research on magpies, road ecology and the effects of bird feeding. Today, he is the author of eight successful books about birds and road ecology[14].

WANTED

Professor Darryl Jones

aka the Turkey Whisperer

CATCHING A TURKEY 1998   WHISPERING TO TURKEY CHICK   STUDYING TURKEY MOUND

Professor Darryl Jones is the first
non-Indigenous person in Australia to have studied
brush-turkeys in detail, since 1983.
He is now wanted for a

Brush-turkey Interview

Last seen: Kuala Lumpur, writing successful books on
birds.

Previous haunts: Griffith University, Brisbane

Occasionally spotted watching rugby games

A$10,0000

While Darryl Jones focused on the behaviour and ecology of the BT, a second group of researchers started in 1987 to study the physiological adaptations of BTs, meaning the internal functions that help these birds cope with the unusual breeding behaviour. These scientists from the Universities of Queensland and Adelaide (Roger Seymour, David Booth and colleagues) have provided fascinating insights into how eggs and embryos cope with being stuck in the soil, summarised in Chapters 12 and 14 of this book.

Darryl not only supported me with my research work on BTs, but also three other Honours or PhD students who chose the BT for their postgraduate studies: Kate Barry, David Wells and Matthew Hall. Matthew is also part of a research group investigating how and why BTs spread through Sydney (see Chapter 18). These BT researchers are aided by an army of citizen scientists who report sightings in the Big City Birds phone app and use social media campaigns to highlight urban BTs.

Scientists from overseas also show great interest in BT research. One such study focused on analysing videos of how the chicks move compared to other bird hatchlings, or on the amazing ways the eggs are guarded against the invasion of bacteria in the soil (see Chapter 14). Several international researchers cooperated with me to study the behavioural development and egg yolk composition of BTs (see Chapters 13 & 14), and a current research project by Jingmai O'Connor from the Chicago Field Museum of Natural History looks into the microscopic structure of bone tissues (histology) compared to other birds.

Compared to how much time has passed since the first formal studies in the 1960s, only relatively few researchers have worked on BTs. I encourage students and researchers of the future to take up the challenge and discover more about this representative of an ancient avian line in Australia, as there is still so much to learn.

Many Australians are hobby birdwatchers and love observing these birds. They may even be surprised that some of what they have witnessed has not yet been described by science. There are also all those town folk who show great pride in their

BTs, such as in Noosa (Queensland), where a white BT became a celebrity (see Chapter 2) and in Milla Milla (Queensland), where the BTs feature prominently on the entrance sign to the little town. Australia Post recently acknowledged the public's interest in megapodes, including the BT, by releasing a beautiful set of stamps to celebrate them. Artists all over Australia incorporate BT designs creatively on t-shirts, jewellery, paintings, sculptures, letter weights and even as applications on underwear. I have been presented with BT-shaped cookies and cakes — I'm always amazed at the new ways people use BT shapes in their designs.

Do the above art and craft items indicate how many people like BTs? A survey in Queensland found that about 75% of people have come to admire or at least tolerate these birds. And to my surprise, the BT scored high in the 2021 *Guardian Bird of the Year Poll*. Despite not being as Instagrammable as the top four birds, it was the fifth most popular bird for thousands of voters in this poll!

Yet many Australians are unwilling to share their dwellings with big birds that cause a substantial mess, destroy gardens and cause chaos of various sorts. How can you deter BTs from your property?

Section 4
Living in Harmony
CHAPTERS 17-18

Top: A large brush-turkey mound fenced in to (somewhat) contain the 'mess'.
Photo: D. Booth

Left: Some people accept or even welcome brush-turkeys near their homes.
Photo: E. Mann.

17

How can I deter them?

There are a few things you can do

Stop attracting them to your backyard

Composting your food waste is a great way to live more sustainably but, if you don't want BTs in your backyard, cover up that compost heap. They love to take food scraps from it! Talking about accidental feeding, you also need to ensure you don't provide other types of food. Bird feeders are a no-no if you are serious about deterring BTs, as they love birdseed from the feeder or on the ground underneath. Pet food, such as dog and cat biscuits or tinned food, is also a delicacy for them. So don't feed your pets outside the house, even if that increases the mess inside. BTs will scare your cat or small to medium-sized dog away from the bowl outside so they can indulge in that nutrient-rich food. If the dog is larger, they'll hang around till it has left to peck up the last scraps they can find.

If you have piles of spare mulch or leaves lying around your backyard, you will attract male BTs who may regard this as the perfect start of an incubation mound. Keep your yard raked free of leaf litter and cover up any extra piles of mulch or leaves with heavy tarpaulins if you don't want a new BT nest.

A more costly method of deterring BTs is investing in a Garden Sentinel or Sonic Scarecrow, an automated sprinkler triggered by an infrared heat sensor that sprays water onto the birds each time they walk by — a bit like an automated water pistol. You can find various suppliers online, and they cost between $150 to $200 in Australia. I have had reports of these devices successfully deterring BTs, especially in the Brisbane area.

How to protect your plants

Many people living with BTs successfully grow flowers, vegetables and other more fragile plants by inserting vertical sticks, such as the branches of eucalyptus trees, in the ground around the plants to form barriers. Large logs or small rocks placed next to, or near, the plants may also do the job, as does covering the area with large, heavily branched sticks and tree loppings. The idea is to prevent the BTs from scratching in those areas or at least make it difficult for them. The same can be achieved by laying wood pallets from a building supplier on the ground and placing the plants in the gaps between the wood.

If the above methods don't work, you may have to invest more time and money. One option is to put down matting that stops the birds from scratching and secure it with stakes or logs on top of the soil and mulch, or place it underneath your mulch. You can purchase this from hardware stores as so-called mulch matting, and it is made from either coir, jute or polyester (less suitable for eco-conscious people). Some people also use chicken wire and report that the BTs abandoned their place because they did not like their feet getting caught in it. Wire, however, is more likely to cause injury to the BTs' feet and claws.

Some gardeners invest in special plant protectors, such as individual tree guards or Vegepods (raised and enclosed garden bed kits of various sizes). Others build their own raised garden bed. The BTs don't usually jump up into raised garden beds that are at least 70 centimetres (27 inches) off the ground, even if it is not covered. This is what I have been told, but I do not take the blame if the occasional BT does figure out how to reach those raised plants. If they do, you may need to cover them with a shade cloth structure (such as a Vegepod cover) or different types of netting with small holes to ensure other wildlife such as bats and birds don't get entangled. Enclosed garden beds also have the additional benefit of keeping out hungry possums at night. Once your plants or seedlings are strong and tall enough to withstand a turkey attack, they may no longer need extra protection.

When planting out your garden, you may want to consider two approaches that help minimise damage from BT raking. One is to plant in stages instead of all at once so that you can pay more attention to protecting individual plants while they are still small and more fragile. The other is to coincide your planting with the non-breeding season (February to May, June or July, see Chapter 9), especially if you have a mound nearby. The male's urge to rake is not so great during this period so he does less damage to your plants. You should also not place any new plants near a dormant existing BT mound unless you plan to get rid of the mound during the non-breeding season (see below).

Choice of plants

You may have some areas in your garden where you would like to plant flowering groundcover. Many native plants provide thick groundcover with flowers. Examples include plants from the prostrate grevillea family, clumping lomandra and dianellagrasses. Bromeliadsplanted close to each other also create a thick groundcover. BTs are unlikely to rake these types of foliage. Low-growing plants also provide suitable habitat and cover for other native animals, such as small birds, reptiles, small mammals and invertebrates.

When you first put young plants in the ground, you may want to consider using landscaping materials between them in the spaces you leave for the plants to grow. As mentioned above, use heavier materials such as pebbles or river rocks. Ask your nursery or landscape supplier and they can show you a range of plants and materials to choose from.

How to deter mound-building or move a dormant mound

How can you stop BTs from building a mound? Advice number one is to be watchful early in the breeding season, which is July/August in Sydney and May/June further north. This is when the males start building their mounds and it is the only time you can still remove these incubators. You can get rid of the leaf litter they are accumulating and this will hopefully deter them, but you may have to do it a few times before they give up. If you spread out the material they will likely rake it all together again, so it must be entirely removed from your yard. Alternatively, you can cover a mound that does not yet contain eggs with a heavy-duty tarp, black plastic or shade cloth to deter the male.

You can also make the male's preferred location for his mound less attractive by selectively pruning some of the shade-producing vegetation above the mound. The nests require about 80% to 90% shade otherwise the incubation temperatures inside will rise too much under the glaring sun. If you remove the shade cover, the male may move elsewhere. Pruning a few branches is usually permissible but if you want to cut down the whole tree, you may need to check if you need a council permit for that.

Many people don't act early when the male starts building his incubator. They find it fascinating to watch this action for a while until they realise the mess it makes. And then it's too late. The male will be so determined to build that no matter what you do, he will keep raking together whatever he finds. Soon, removing the mound will be illegal because the females will have laid eggs into it. Removing a mound with eggs inside harms native wildlife and incurs heavy penalties (see below).

If you have a large property, there might be a spot where you wouldn't mind the male building instead of right next to your access path, vegetable garden or flower beds. You can try attracting this eager architect to that more suitable spot by starting a mound of leaf litter, mulch or lawn clippings, or a combination of all three. Make sure it is in a partly shady spot beneath a tree. The males are known to take over such existing mounds as it means less work for them, and you benefit by having lured them to a more acceptable location.

Once the breeding season ends around late January and there are clearly no more eggs hatching, it is legal once again to remove the mound. However, even if the male has stopped working on his mound, there may still be chicks hatching from it for up to seven weeks (the incubation time is 49 days). We don't know whether the male stops working once all the chicks have hatched or if he abandons his mound beforehand. To be safe and avoid penalties, it is advisable to wait for about seven weeks after you last saw the male working on the mound. Apart from the male's absence, another tell-tale sign of an abandoned mound is if you see small plant seedlings growing on top of the mound and if the area nearby is no longer raked clean of any loose material. Then you may safely remove the mound, and many garden owners welcome the excellent fertile soil to spread over their garden.

Doubtful methods

If you have desperately scoured the internet for BT deterrents, you may have come across the idea of placing a large mirror strategically near the male's mound. He will perceive his mirror image as another male in his territory and attack it. The idea is that he will move on after some days because he tired of a rival in his territory. Unfortunately, I have seen the opposite: the male kept attacking the mirror for many days and ended up with a severely wounded head. The skin on the BT's head is relatively thin and protected by only a few stubble feathers, and a male that constantly runs into a mirror soon ends up with head injuries. This is not an ethical way of deterring BTs.

The teddy bear method is also circulating on the internet. It is amusing but most likely a waste of money – and teddy bears. One garden owner swears that she deterred BTs from her garden by attaching teddy bears to stakes positioned all around her backyard. She claimed that the big eyes of the bears averted the BTs. Another desperate gardener, scientist Lesley Hughes, followed her method but then reported: "*The teddies were duly staked and positioned, the backyard now the scene of some bizarre druidic sacrifice staged by toddlers. By the next day, two of the teddies had been interred within the mound*". Lesley tried several other methods, but none worked, and she eventually gave up and handed over her backyard to the BT[1].

A third method with mixed results involves applying a strong-smelling substance to your garden, assuming the smell will deter BTs. This could, for example, be Dynamic Lifter with its strong and offensive stench. It may deter the birds for a few weeks, but after that, when the Lifter has broken down and no longer giving off bad smells, they are likely to return.

The no-no's and legal side of things

All methods to deter BTs must be harmless as they are a native species protected by law. In Queensland, under the Nature Conservation Act 1992, it is an offence to harm BTs. Penalties for taking or killing a BT range from $667 to a maximum of $133,000 or up to two years in prison. In New South Wales, harming a BT can result in fines of up to $22,000 under the Biodiversity Conservation Act 2016.

The following activities can legally be considered harm to BTs:

- Killing or injuring them (such as by using traps, poison or weapons);

- Catching them, with or without translocation to another area;

- Destroying mounds that contain eggs;

- Fencing in mounds that contain eggs because these eggs are destined to die if the male can no longer tend the mound; and

- Removing and destroying eggs from mounds.

Some BT haters nevertheless try to remove the birds by harming them, thinking that authorities won't care about such a common species. But a recent case in Sydney shows they are wrong. Someone poisoned BTs in Mosman and the media and authorities paid a lot of attention to this case, working hard to identify the culprit[2].

INTERVIEW

HOLLY PARSONS, BIRD LIFE AUSTRALIA

WHAT IS YOUR ROLE?

66 I am the Urban Bird Program Manager at BirdLife Australia, Australia's largest bird conservation charity. We are dedicated to the conservation and protection of Australian native bird species and their habitats through research, advocacy, education, and community engagement. 99

HOW CAN WE HELP BRUSH-TURKEYS IN OUR YARDS?

66 Living harmoniously with brush-turkeys is not only about protecting your garden but also appreciating the unique presence of these fascinating native birds. Brush-turkeys are a remarkable part of Australia's wildlife, and understanding their value can help create a more peaceful coexistence. With a few strategies, you can minimise potential conflicts and enjoy their presence.

First, it's essential to understand their behaviour. Brush-turkeys are curious and often visit gardens in search of food and suitable sites for mounds. Knowing this can help you anticipate their visits, plan for their arrival and make necessary adjustments. They do play an important role in our ecosystems - even in the ones in our gardens - by acting as a natural pest control, aerating the soil and dispersing seeds.

The biggest conflict comes when a male builds a mound in your yard, in a space where you don't want him to do so. Start off on the right foot by securing compost heaps and garden beds. Put tarpaulins over compost heaps or mulch piles. These birds are notorious for scratching and digging, so use wire mesh (like chicken wire) just under the top of the mulch or compost, or use fencing, tree guards or small rocks to protect your plants and the base of trees.

INTERVIEW HOLLY PARSONS CONTINUED

This not only safeguards those features of your garden but can also direct them towards an area of the yard you are happy with them using. They like shady areas for their mounds, so trim some tree branches in areas you don't want them digging in and offer an open mulch or compost pile in a shady area under some trees where the male may move to for mound building. Provide him with his own mulch supply nearby, and he will be very appreciative.

Once a mound is established, there is little to do to stop the male from carrying out his duties (and harming him or the eggs is illegal) - so it is time to sit back and enjoy him and his antics. Keep cats indoors and supervise dogs as the young are very susceptible to predation - they are on their own, so we need to give them a fighting chance to make it in their life in the suburbs. 🙙

Calling the pest controller

Local pest controllers must show a valid licence, such as a 'Licence to Harm Native Animals' from the NSW National Parks and Wildlife Service[3] or a 'Wildlife Movement Permit' from the Queensland Government[4]. This entitles them to remove a BT from your property without acting illegally. They try to capture the birds without harming them and then usually relocate them to another area. Such translocation is rarely successful, however, and often leads to the death of the bird as it follows its instinct to return to its territory.

A ranger from the NSW National Parks and Wildlife Service and I translocated two adult males to a location north of Sydney, 100 kilometres (62 miles) from where they had been caught. We monitored them via radio transmitters and it only took two days before we found both of them dead on a major highway because they had tried to return home and were hit by cars while crossing the busy road. A translocation is often a death sentence. The area where the birds are being moved to is usually already occupied by another male who will violently expel the newcomer. In addition, another BT will likely move in and replace the one removed unless you make the place unattractive for BTs in the meantime.

Unfortunately, there is no guaranteed method for deterring BTs. The above methods have worked for many people, but others have found some middle ground, a way to co-exist with these birds. The birds are native and they are here to stay, and I can only hope that this book helps you better understand how BTs tick and why they do what they do.

Figure 26. Don't let that turkey get you down (old English saying). Drawing Ann Göth, inspired by Sandra Boynton's original cartoon[5].

18

How can I help?

What to do with injured or 'abandoned' brush-turkeys

Reporting injured or 'abandoned' BTs

Many volunteers working for wildlife rescue organisations on Australia's East Coast receive all-too-regular calls from members of the public reporting an abandoned brown fluffy baby bird needing immediate help. By now, you know that this little feather-ball is often not helpless but an independent BT chick that can perfectly fend for itself. They usually don't need our care, and rescue organisations would appreciate it if more people knew not to block up their helplines with calls about BT chicks unless they appear injured or unwell.

Another regular call to wildlife rescue hotlines comes from well-meaning people who observe an adult BT with an injured leg. The interview with Bev Young OAM from Sydney Wildlife Rescue (see Chapter 3) tells us that often it is better to wait before intervening because the birds cope quite well in adapting to their leg injuries. As long as the BT is observed still feeding itself, it is better to leave it alone. We have heard of several instances where a limping BT healed itself and walked normally again after some weeks. However, a call to the hotline is warranted if there is a serious threat to a BT's wellbeing or if the animal is suffering from bad injuries and requires euthanasia.

Feeding BTs

What about feeding BTs? You may mean well, but by feeding them you cause more harm than good. Your food may not match the variety of food these birds consume naturally and it may make them sick. Also, the population of local BTs will quickly increase and reach a size that it would not reach in a natural situation without additional food.

In metropolitan areas, large flocks of BTs cause more conflicts with people, eventually leading to more calls to get rid of them. In bushland areas, a high population of BTs causes a significant disturbance to the environment because too many individuals scratching for food disturbs the soil so much that new saplings cannot grow. As a result, there will be fewer plants covering the ground, leading to soil erosion[1].

While supplementary food is undesirable, you may like to provide a bowl of drinking water on the ground especially during droughts and heat waves. The BTs along with all the other birds and animals visiting your property will appreciate it! Remember to place a tall rock in your bowl to allow smaller birds to find refuge if they become waterlogged while bathing, and to change your bowl's water frequently.

Record your sightings

You may also consider contributing to the study of BTs by entering your sightings into the Big City Birds app[2]. The app allows you to record the BTs' whereabouts and whether they have a nest or carry a numbered tag on their wing. This helps researchers understand how BTs adapt to the challenges and opportunities in a big city and gives us a better idea of their territory and range. Matthew Hall gives a more detailed description in his interview below.

INTERVIEW

MATT HALL, THE UNIVERSITY OF SYDNEY

WHAT IS YOUR ROLE?

66 I am an ecologist. In 2022, I completed my PhD on urban brush-turkeys at The University of Sydney. 99

WHAT IS THE BIG CITY BIRDS APP GOOD FOR AND HOW CAN PEOPLE USE IT?

66 The Big City Birds app is a citizen science project developed by Spotteron and jointly managed by The University of Sydney, The Max Planck Institute of Animal Behavior, and the Australian National University. Developed initially to collect community sightings of brush-turkeys, the project has expanded to focus on five iconic urban bird species: the sulphur-crested cockatoo, Australian white ibis, little corella, long-billed corella, and, of course, the Australian brush-turkey. While these bird species are the main focus of research, citizen scientists can make a report about any bird they see.

The project aims to engage the general public and keen bird watchers to report sightings of these birds, including how they behave in both urban and natural habitats. The data collected will help researchers understand how urbanisation affects the behaviour, movement, distribution, and reproduction of these species and identify the traits that have allowed these species to adapt to big city life. 99

<u>Welcoming them to your garden</u>

If a BT has decided to build a mound in your garden and you welcome him, there are a few things you can do to make his life easier. You may want to provide him with his own mulch supply, which also takes the pressure off your garden. Perhaps a friend or neighbour has an extra supply of leaf litter or wood chips they no longer require? He also needs a certain degree of moisture in his mound, so you could sprinkle the mound base on hot and dry days with a bit of water. Just be careful it doesn't become waterlogged. The male will decide how much of this moist material he will work into the mound's core, where the eggs are incubated.

Once chicks hatch from the mound, you can best help them survive by keeping away any cats and by providing coverage under which they can hide. This could be a temporary cover or some ground-covering plants that are a more permanent feature of your garden. We know thickets considerably increase the chicks' survival chances and that many of them are eaten by cats, kookaburras or crows (see Chapter 13). There is no guarantee that the chicks will stay in your garden as they sometimes migrate far, but the greater the coverage as well as the natural food supply in your soil and mulch, the more likely they will hang around for a while as they develop.

<u>Spread the word</u>

The most important contribution you can make to the continuous survival of our iconic representatives of an ancient avian line is to spread the word. The BT conflict in some of our East Coast cities and towns is unique to Australia. Understandably, some people don't want car-sized mounds and messes on their property. At the same time, we must accept that these large native birds were here first and are here to stay. The ancestors of BTs first roamed this country at least three million years ago. You can spread the word that BTs belong to Australia. You can pass on some of the tips on how to deter BTs from your garden (see Chapter

17) to those who either don't want BTs altogether or who need to find a way to live more harmoniously with them.

Australia's East Coast is full of birds that we perceive as inconveniences. Ibises spread our garbage far and wide, cockatoos chew our balcony railings or antennas to bits, magpies attack and inflict serious injuries and the curlews' nightly haunting calls trigger false police alerts. Not to mention the koel and other noisy big birds that rob us of our early morning sleep! BTs also cause an inconvenience, yet some people find it harder to accept these troublemakers than they do the other 'nuisance' birds. This is partly because many East Coasters know little about them. You can change other people's views of BTs by passing on what you learned from this book.

If you are the manager of a property where a BT has built his mound, such as for an owners' co-op or strata management, you could help educate people by erecting signs that explain why these birds are causing a mess — see the sign shown here as an example.

I hope that after reading this book you now appreciate BTs as the iconic Australian wildlife they are, or that you have at least enough information on how to live more harmoniously with these annoying and amazing native birds. After all, international bird watchers from around the world travel all the way to Australia to observe the BTs extraordinary behaviour, and the souvenir shops at Sydney airport now stock BT soft toys for tourists to take home as a souvenir. Hopefully more Australians will also appreciate the opportunity to watch their unique nesting, social and chick behaviour that is not found in any other birds apart from their few megapode relatives. And best of all, you don't need binocu-

lars or to crane your neck — it's all happening right there at ground level, possibly in your own backyard!

And if you are a BT fan, you may also like to pay more attention to the plight of its cousin, the malleefowl. This species shares the BT's unique approach to breeding and child-rearing but, unfortunately, is highly endangered and needs our help. Head to the website of the National Malleefowl Recovery Group to find out more (www.nationalmalleefowl.com.au).

ACKNOWLEDGEMENTS

I thank my son Toby Taylor who, in a random conversation about which book I should write next, suggested that this is the obvious one. People tell me he had the right idea, and I don't know why I didn't think of it myself. I am also grateful to Professor Darryl Jones for encouraging the writing of this book, as it is really him, the 'brush-turkey whisperer', who should have written it. Without Darryl and his tireless support, I would have never come to Australia to study these amazing annoying birds.

It is hard to thank everyone who assisted me in my many years of brush-turkey research, and my apologies in advance if I have forgotten some of you. Chris Evans and Marielle Herberstein enabled me to do postdoctoral research on brush-turkeys at Macquarie University. Without my student volunteers and other helpers, I could not have dug through so many mounds by hand and spent so many hours watching chicks in digging boxes in a dark room. I know these helpers would have done this for me even if I had not fed them chocolate bars! Several scientists cooperated with me on different research projects, and I hope I have mentioned them all in this book. The numerous landholders who gave me access to their property for my studies of eggs, chicks, or adults often did so with open arms and an open mind to learn more about these birds. I also thank the many people who shared their amazing knowledge about, and observations of, brush-turkeys with me. These include Indigenous Australians (especially Dr Shane Smithers and Warren Canendo), hobby birdwatchers, attendees at my talks, land managers, rangers and scientists who taught me new methods for my studies. Some of these people volunteered to be interviewed for this book and I am grateful for their time, especially as we met during the brush-turkey breeding season, when they were more than busy responding to calls from residents complaining about annoying brush-turkeys.

Writing and illustrating this book was enjoyable, and I thank my advanced readers who gave early feedback and improved the first draft, especially David Booth, René Dekker, Cliff Frith OAM, Matt Hall, Judy Harrington, Roger Seymour and Bev Young OAM. My husband Wayne deserves a big thanks for his constant support and for listening to more than his fair share of brush-turkey book talk. Emily Mann, my editor, seems to almost love brush-turkeys more than me and did an amazing job smoothing out my English and making sure it all made sense. The members of my Facebook "We Love Memoirs" authors' group provided valuable tips for the cover design and self-publishing process. And last but not least, thank you to my readers who are willing to learn more about brush-turkeys, in some way or another, and have taken an interest in this book.

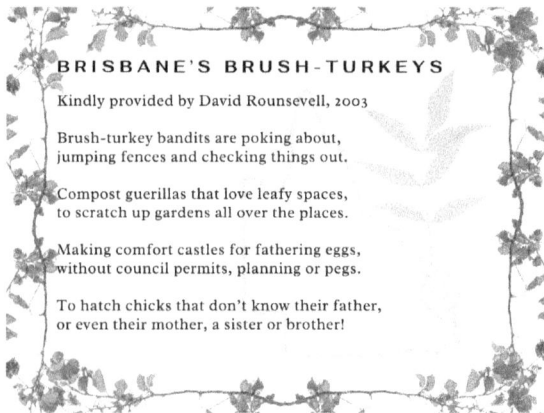

BRISBANE'S BRUSH-TURKEYS

Kindly provided by David Rounsevell, 2003

Brush-turkey bandits are poking about,
jumping fences and checking things out.

Compost guerillas that love leafy spaces,
to scratch up gardens all over the places.

Making comfort castles for fathering eggs,
without council permits, planning or pegs.

To hatch chicks that don't know their father,
or even their mother, a sister or brother!

Figure 27. A poem for Brisbane's Brush-turkeys. Written by David Rounsevell, 2003, and reproduced with his permission.

Figure 28. Linocut of a Brush-turkey, created by Rex Addison as an illustration for the book "Mother Lode: Stories of Home Life and Home Death" by Susan Addison. University of Queensland Press, 2001. Reproduced with permission from Rex Addison.

ABOUT THE AUTHOR

Dr Ann Göth is a science teacher, ecologist, researcher, academic, speaker and writer based in Sydney. She followed Professor Darryl Jones's footsteps to become one of the first Australian scientists to study brush-turkeys. These birds were the reason she moved from Austria to Australia, and they have been a big part of her life ever since. Ann researched BTs, especially their chicks, for her PhD and postdoctoral studies. In 2008, she co-wrote the book *Mound-builders* with Darryl Jones about brush-turkeys and their two Australian megapode relatives. Ann is often asked to contribute her insights into brush-turkey behaviour and ecology for magazines, radio interviews, local council presentations and TV shows. She enjoys giving public presentations for Sydney councils on how to live more harmoniously with urban brush-turkeys that are big, amazing, annoying and often little understood.

Check out her website https://anngothauthor.com

MORE BOOKS BY ANN

BUSH-TURKEY NEEDS A FRIEND
Illustrated and written by Ann, for
kids aged 3-8

Embark on a delightful animal adventure with the charismatic Tom Turkey in 'Bush-Turkey Needs a Friend'. This enchanting children's book brings to life the whimsical tale of Tom as he navigates the challenges of growing up all on his own.

Follow Tom's quest to meet other birds that could be his friends, only to find out that he is different to all of them. But when he grows up and builds a huge, strange-looking nest of leaf litter, he eventually discovers that finding friends can take time and that being different makes you unique.

With captivating illustrations and a delightful storyline, this book for children aged 3-8 captures the magic of friendship and liking yourself even if you are different. 'Bush-Turkey Needs a Friend' will teach valuable life lessons and give young readers a newfound understanding of the uniqueness of the Brush (Bush) Turkey, a remarkable Australian native bird found in bush and backyards.

Published November 2023. Available on Amazon or order from your nearest bookstore ISBN 978-0-6486037-2-6

VOLCANIC ADVENTURES IN TONGA

Follow Ann as she studies a relative of the Brush Turkey on a remote island in Tonga! In this travel memoir, Ann takes you on a rollicking adventure to experience authentic Polynesian culture and the unique adaptations of a volcano-breeding bird on one of the most remote Pacific islands imaginable. She quenches the thirst of bird and nature lovers as well as travellers with a wanderlust for faraway islands while taking you to volcanoes overdue to erupt and coral cays rarely visited by tourists.

This narrative shows what it can be like to live a simple existence on a remote island, and demonstrates the good and bad sides of 'Tongan Time', that is, in a nutshell, the concept that you enjoy the presence and don't worry about the future. It leads you to be fascinated by birds that use a volcano as an incubator and to be pondering the resilience needed when confronted with cyclones, paucity of fresh food, deteriorating equipment, stinging bugs, oppressive humidity, illness and the threat of being stuck on an island for much longer than expected. The story is underpinned by letters Ann wrote during a 17-month conservation project for the endangered Polynesian Megapode on Niuafo'ou (Tin Can Island) in Tonga, 30 years ago.

Published March 2023. ISBN 978-1035809516

This book is available as eBook or paperback, details on https://anngothauthor.com. Available from Amazon, Barnes & Noble and several other bookstores.

MOUND-BUILDERS. 2008. Darryl Jones & Ann Göth.

Mound-builders are unique in being the only birds that do not incubate their eggs using body heat; rather, a variety of naturally occurring sources of heat is exploited such as solar energy and the heat generated by decomposing organic matter. This book shows how this remarkable adaptation influences

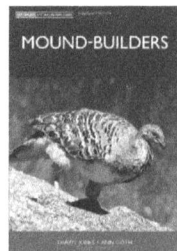

every part of these birds' lives, including the development of the embryo, the parentless life of the hatchlings, their social organisation and their survival.

Twenty-two species of mound-builders exist within the Megapode family. *Mound-builders* examines the three occurring in Australia: the scrubfowl in the humid tropics; the brush-turkey in dense forested areas from Cape York to Sydney; and most remarkable of all, the malleefowl in the arid interior.

Scientific interest in these birds has increased considerably in recent decades, and *Mound-builders* summarises many significant discoveries. With a strong emphasis on conservation and changing interactions between mound-builders and people, this is an excellent introduction to one of the most unusual bird families.

Available as eBook only, on Amazon. Paperback out of print. ISBN 978-0643093454

Thank you for reading my book! I am grateful and hope you found value in it. Please consider sharing it with family or friends and leaving a review online, as this would be very valuable in my quest to promote this book to a wider readership.

Would you like to see
brush-turkey photos?

To reduce printing costs and keep
this book affordable, it does not
contain colour photos of
brush-turkeys.

BUT

You can scan this QR code to view
colour photos that show some of the
features and behaviours mentioned
in this book.

ENDNOTES

Introduction: Urban conflicts with unusual birds

1. Rickards, B. (2011). Talking turkey – and wildlife corridors. *Eco*, April/May 2001, 11 & 21.

2. Taylor, J. (2013). *Brisbane's brush turkey explosion heads south*. ABC News. https://www.abc.net.au/n ews/2014-12-26/scientist-cant-explain-brisbane-brush-turkey-explosion/5987606 AND Jones, D.N. & Everding, S.E. (1991). Australian Brush-turkeys in a suburban environment: implications for conflict and conservation. *Wildlife Research*, 18, 285-297.

Common complaints

1. https://berowrabackyard.com/category/native-animals/brush-turkeys/

2. Dani, J. (2021). *Brush Turkrey v Lace Monitor*. [YouTube Video]. https://www.youtube.com/watch v=J4jom9HjMmc&ab_channel=JohnDani

3. StoryfulViral. (2021). *Australian Brushturkey Tires to Take Dog for Walk in Sydney*. https://www.y outube.com/watch?v=zA-N-PnUSR8&ab_channel=StoryfulViral

Brush-turkeys in the media

1. McGregor, A. (2016). The March of the Brush-turkey. *Australian Geographic*, Sept 22, 2016. 45-49.

2. https://www.abc.net.au/news/2016-03-31/brush-turkeys-haunting-sydney-backyards/7287518

3. https://www.smh.com.au/national/nsw/never-seen-that-before-brush-turkeys-are-turning-carnivoro us-in-sydney-suburbs-20220603-p5aqt5.html

4. https://thenewdaily.com.au/news/state/nsw/2015/10/06/brush-turkeys-invading-sydney/

5. https://www.dailytelegraph.com.au/news/nsw/tweed-heads/are-bush-turkeys-becoming-a-plague/ne ws-story/5819cbdb7d5bc53616994730534fedc1

6. https://www.abc.net.au/news/2009-08-17/man-v-bird-the-brush-turkey-battle/1394040

7. https://www.dailytelegraph.com.au/newslocal/north-shore/brush-turkeys-are-back-in-force-and-driv ing-north-sydney-gardeners-mad-with-their-backyard-antics/news-story/beb742d047ca803c8f1a1fc1 606be64b

8. https://www.abc.net.au/news/2009-07-08/back-from-the-bush-turkeys-hit-sydney-backyards/13461 72

9. https://www.dailytelegraph.com.au/news/nsw/lismore/dealing-with-brush-turkeys-the-bird-with-the-bad-reputation/news-story/7e8834ddbecc3a260c3f11f1ca0bfcc9

10. https://www.abc.net.au/news/2018-10-27/brush-turkeys-are-booming-in-urban-areas-and-we-dont-know-why/10406294

11. https://www.abc.net.au/news/2022-08-03/brush-turkeys-reclaim-sydney-south-inner-city-sightings/101291730

12. https://www.smh.com.au/national/nsw/why-brush-turkeys-are-headed-to-a-sydney-suburb-near-you-20180102-h0cc65.html

13. https://www.news.com.au/technology/science/animals/brush-turkeys-are-on-the-march-colonising-our-inner-cities/news-story/490add8c0dcfcace10bc24c145757510

14. https://www.abc.net.au/news/science/2017-01-17/five-reasons-to-love-brush-turkeys/7199724

15. https://www.smh.com.au/national/nsw/fowl-play-who-s-poisoning-brush-turkeys-on-sydney-s-lower-north-shore-20230926-p5e7ot.html

16. https://www.abc.net.au/news/2018-11-08/brush-turkey-arrow-attack-in-the-suburbs/10467510

17. ABC Radio Sydney. Brush-turkey's story of survival. ABC Radio Sydney. Broadcast Wed 27 Sep 2023 at 8:30 am. https://www.abc.net.au/listen/programs/sydney-mornings/jmartin/102907264

18. https://www.smh.com.au/national/nsw/basil-the-bandit-has-me-googling-brush-turkey-recipes-20210329-p57exo.html

19. https://www.smh.com.au/national/nsw/keep-basil-the-brush-turkey-off-the-menu-20210405-p57gl3.html

20. https://www.dailymail.co.uk/femail/article-5006965/Couples-wedding-photos-interrupted-brush-turkey.html

Are they intruding or invading?

1. ABC News. (2017). Giant brush-turkeys roamed the Australian landscape. https://www.abc.net.au/news/science/2017-06-14/giant-brush-turkeys-roamed-the-australian-landscape/8613944

2. Boles, W.E., (2008). Systematics of the fossil Australian giant megapodes *Progura* (Aves: Megapodiidae). *Oryctos*, 7, 195-215.

3. https://theconversation.com/tall-turkeys-and-nuggety-chickens-large-megapode-birds-once-lived-across-australia-79111

4. Göth, A., Nicol, K.P., Ross, G. & Shields, J.J. (2006). Present and past distribution of Australian Brush-turkeys *Alectura lathami* in New South Wales – implications for management. *Pacific Conservation Biology*, 12: 22-30.

5. Hall, M., Martin, J., Burns, A. & Hochuli, Dieter. (2021). The Decline, Fall, and Rise of a Large Urban Colonising Bird. 10.21203/rs.3.rs-1179603/v1.

6. https://bie.ala.org.au/species/https:/biodiversity.org.au/afd/taxa/3c6fe522-90df-4447-ab21-1dd086 461abc

7. McGregor, A. (2016). The March of the Brush-turkey. *Australian Geographic*, Sept 22, 2016. 45-49. AND 9 News Australia. (2023). *Brush turkeys spreading through south of Sydney.* https://www.you tube.com/watch?v=Q46Jy1WOcR0&feature=youtu.be&ab_channel=9NewsAustralia

Why are they returning now?

1. Sydney Living Museums. (2018). Let's talk turkey – brush style. https://blogs.sydneylivingmuseum s.com.au/cook/lets-talk-turkey-brush-style/

2. *Hannah Maclurcan, Mrs Maclurcan's Cookery Book. 1903 p. 214. https://ia902203.us.archive.org/21/items/b20417470/b20417470.pdf*

3. Hochuli, D. (2023). Our cities can become safe havens for wildlife [Podcast]. Australian Geographic. https://podcasts.apple.com/au/podcast/talking-australia/id1464668928?i=1000466036347

4. Hall, M.J., Martin, J.M., Burns, A.L., and Hochuli, D. F. (2022). Unexpected dispersal of Australian brush turkeys (*Alectura lathami*) in an urban landscape. *Austral Ecology*, 47, 1544-1548.

5. Göth, A. & Vogel, U. (2003). Juvenile dispersal and habitat selectivity in the megapode *Alectura lathami* (Australian brush-turkey). *Wildlife Research*, 30, 69-74.

What do they look like?

1. Jones, D.N., Dekker, R.W.R.J. & Roselaar, C.S. (1995). *The Megapodes: Megapodiidae.* Oxford University Press.

2. same as last endnote

3. Darryl Jones, personal communication.

4. Jones, D.N. (1990). Social organization and sexual interactions in Australian Brush-turkeys (*Alectura lathami*: implications for promiscuity in a mound-building megapode. *Ethology,* 84, 89-104.

5. Olsen, V. (2002). Evolution of Avian Carotenoid pigmentation: behavioural, biochemical and comparative approaches. PhD thesis, University of Queensland, Brisbane.

6. ABC News (2022). *Rare white brush turkey in Noosa amazes scientists as species boom in urban areas.* https://www.abc.net.au/news/2020-03-07/albino-brush-turkey-noosa/12024526

7. Jones, D.N. & Göth, A. (2008). *Mound-builders.* CSIRO Publishing.

Why the mound?

1. Jones, D.N. (1988). Construction and maintenance of the incubation mounds of the Australian Brush-turkey *Alectura ltahami*. *Emu,* 88, 210-218.

2. Seymour R.S. & Bradford, D.F. (1992). Temperature Regulation in the Incubation Mounds of the Australian Brush-Turkey. *The Condor*, 94(1), 134-150.

3. Jones, D.N. (1988). Hatching success of the Australian Brush-turkey *Alectura lathami* in Southern Queensland. *Emu,* 88, 260-263.

4. Harris, R.B., Birks, S.M. & Leaché, A.D. (2014). Incubator birds: biogeographical origins and evolution of underground nesting in megapodes (Galliformes: Megapodiidae). *Journal of Biogeography*, 41, 2045-2056.

Mound-building - How do they do it?

1. Jones, D.N. (1988). Construction and maintenance of the incubation mounds of the Australian Brush-turkey. *Emu*, 88, 210-218.

2. Same as last endnote.

3. Jones, D.N. & Göth, A. (2008). *Mound-builders*. CSIRO Publishing.

4. Göth, A. & Astheimer, L. (2006). Development of mound-building in Australian brush-turkeys (*Alectura lathami*): the role of learning, testosterone and body mass. *Australian Journal of Zoology,* 54, 71-78.

5. Jones, D.N. (1987). *Behavioral ecology of reproduction in the Australian Brush-turkey Alectura lathami*. PhD thesis, Griffith University, Brisbane.

6. Jones, D.N. & Dekker, R.W.R.J. personal communication.

7. Kuenzel, W.J. (2007). Neurobiological basis of sensory perception: welfare implications of beak trimming. *Poultry Science Association*, 86(6), 1273-1282.

8. Frith, H.J. (1962). *The Mallee-fowl: The Bird that Builds an Incubator*. Angus & Robertson, Sydney.

9. Seymour R.S. & Bradford, D.F. (1992). Temperature Regulation in the Incubation Mounds of the Australian Brush-Turkey. *The Condor*, 94(1), 134-150.

10. Göth, A. (2007). Mound and mate choice in a polyandrous megapode: females lay more and larger eggs in nesting mounds with the best incubation temperatures. *The Auk*, 124(1), 253-263.

Brush-turkey's calendar and routine

1. Jones D.N. (1988). Construction and maintenance of the incubation mounds of the Australian Brush-turkey *Alectura lathami*. *Emu*, 88, 210-218.

2. Emily Mann, personal communication.

3. DiBird. (2023). Australian Brush turkey *Alectura lathami*. https://dibird.com/species/australian-brushturkey/

4. Dow, D.D. (1988). Dusting and sunning by Australian Brush-turkeys. *Emu*, 88, 47. doi:10.1071/MU9880047

5. Jones, D.N. & Göth, A. (2008). *Mound-builders*. CSIRO Publishing.

Have a mound? Gotta watch it!

1. Göth, A. (2007). Mound and mate choice in a polyandrous megapode: females lay more and larger eggs in nesting mounds with the best incubation temperatures. *Auk*, 124, 253-263.

2. Jones, D.N. (1988). Hatching success of the Australian Brush-turkey *Alectura lathami* in South-East Queensland. *Emu,* 88, 260-263.

3. Australian Geographic. (2016). *The mystery of brush-turkey sexual aggression* [Video]. https://www.australiangeographic.com.au/topics/wildlife/2016/04/video-brush-turkey-behaviour-and-aggression/

4. Wells, D.A., Jones, D.N., Bulger, D., & Brown, C. (2014). Male brush-turkeys attempt sexual coercion in unusual circumstances. *Behavioural Processes*, *106*, 180–186. https://doi.org/10.1016/j.beproc.2014.06.002 AND Wells, D. (2012). *Mating Behaviour of the Australian Brush-turkey Alectura lathami*. PhD thesis, Macquarie University, Sydney.

5. Birks, S. (1997). Paternity in the Australian Brush-turkey, *Alectura lathami*, a megapode bird with uniparental male care. *Behavioural Ecology*, 8, 560-568.

6. Jones, D.N. (1990). Male mating tactics in a promiscuous megapode: patterns of incubation mound ownership. *Behavioural Ecology*, 1, 107-115. AND Jones D.N. (1988). Construction and maintenance of the incubation mounds of the Australian Brush-turkey *Alectura lathami*. *Emu*, 88, 210-218.

7. Birks, S. (1997). Paternity in the Australian Brush-turkey, *Alectura lathami*, a megapode bird with uniparental male care. *Behavioural Ecology*, 8, 560-568.

8. Jones D.N (1988). Construction and maintenance of the incubation mounds of the Australian Brush-turkey *Alectura lathami*. *Emu*, 88, 210-218.

What are they good for?

1. Hall, M.J. (2022). Ecology of the Australian brush-turkey in Urban Ecosystems. PhD Thesis. University of Sydney. https://hdl.handle.net/2123/29355 AND Hall, M.J., Martin, J.M., Burns, A.L. & Hochuli, D. F. (2023). Mound-building behaviour of a keystone bioturbator alters rates of leaf litter decomposition and movement in urban reserves. *Austral Ecology,* 48(7), 1426-1439.

2. Nugent, D.T., Leonard, S., & Clarke, M.F. (2014). Interactions between the superb lyrebird (*Menura novaehollandiae*) and fire in south-eastern Australia. *Wildlife Research, 41*, 203 - 211. AND Maisey, A.C., Haslem, A., Leonard, S.W.J., and Bennett, A.F. (2021). Foraging by an avian ecosystem engineer extensively modifies the litter and soil layer in forest ecosystems. *Ecological Applications,* 31(1), e0221 9.10.1002/eap.2219

3. The Guardian. (2021). Australian brush turkey sends diamond python running. [Video]. https://www.theguardian.com/environment/video/2021/dec/20/australian-brush-turkey -sends-diamond-python-running-video

4. Göth, A. & Maloney, M. (2012). Powerful Owl preying on an Australian Brush-turkey in Sydney. *Australian Field Ornithology*, 29(2), 102-104.

Chicks covered in soil

1. Göth, A. (2002). Behaviour of Australian Brush-turkey (*Alectura lathami*, Galliformes: Megapodiidae) chicks following underground hatching. *Journal für Ornithologie*, 143, 477-488.

2. Booth, D.T. (1985). Thermoregulation in neonate Brush turkeys (*Alectura lathami*). *Physiological Zoology*, 58, 374-379.

3. Clark, G.A. (1964). Life history and the evolution of megapodes. *The Living Bird, Ithaca, New York*, 3, 374-379.

4. Seymour, R.S. (1984). Patterns of lung aeration in the perinatal period of domestic fowl and Brush Turkey. *In* Seymour, R.S. (ed.) *Respiration and metabolism of embryonic vertebrates*. Dr W. Junk Publishers, Dordrecht/Boston/London. p. 319-332.

Want a self-sufficient kid?

1. Göth, A. & Hauber, M.E. (2004). Ecological approaches to avian recognition systems: benefiting from studies on model and non-model species. *Annales Zoologici Fennici*, 41, 823-842.

2. Göth, A. (2002). Sticking their heads out – life without parents presents many challenges to the brush-turkey chick. *Wingspan*, 12, 8-13. AND Göth, A. (2005). Life without parents. *Nature Australia*, Spring edition, 30-37. AND ABC Catalyst. (2005). *Robo Chick* [TV Documentary]. https ://abc.net.au/catalyst/robo-chick/11008516 AND ABC News. (2004). *Robo-turkey just like the real thing*. https://www.abc.net.au/science/news/enviro/EnviroRepublish_1131408.htm

3. Göth, A. (2001). Innate predator recognition in Australian Brush-turkey hatchlings. *Behaviour*, 138, 117-136.

4. Wong, S. (1999). *Development and behaviour of hatchlings of the Australian Brush-turkey* Alectura lathami. PhD Thesis, Griffith University. https://research-repository.griffith.edu.au/handle/10072/366729?show=full

5. Göth, A. & Evans, C.S. (2004). Social responses without early experience: Australian Brush-turkey chicks use visual cues to aggregate with conspecifics. *Journal of Experimental Biology*, 207, 2199-2208.

6. Barry, K.L. & Goeth, A. (2006). Call recognition in chicks of the Australian brush-turkey (*Alectura lathami*). *Animal Cognition*, 9(1), 47-54.

7. Göth, A. & Proctor, H. (2002). Pecking preferences in hatchlings of the Australian Brush-turkey: the role of food type and colour. *Australian Journal of Zoology*, 50, 93-102.

8. Göth, A. & Evans, C.S. (2006). Life-history and social learning: megapode chicks fail to acquire feeding preferences from conspecifics. *Journal of Comparative Psychology*, 119, 381-386.

9. Göth, A. & Jones, D.N. (2001). Transmitter attachment and its effect on super-precocial Australian Brush-turkey hatchlings. *Wildlife Research*, 28, 1-6.

10. Göth, A. & Vogel, U. (2002). Chick survival in the megapode *Alectura lathami* (Australian Brush-turkey). *Wildlife Research*, 29, 503-511.

11. Göth, A. & Vogel, U. (2003). Juvenile dispersal and habitat selectivity in the megapode *Alectura lathami* (Australian Brush-turkey). *Wildlife Research*, 30, 69-74.

Egg-citing stuff

1. Booth, D.T. (1988). Shell thickness in megapode eggs. *Megapode Newsletter*, 2, 13.

2. Booth, D.T. & Seymour, R.S. (1987). Effect of Eggshell Thinning on Water Vapor Conductance of Malleefowl Eggs. *The Condor*, 89(3), 453-445.

3. Seymour, R.S., Vleck, D., Vleck, C.M. & Booth, D.T. (1987). Water relations of buried eggs of mound building birds. *Journal of Comparative Physiology B*, 157, 413-422. AND Seymour, R.S., Vleck, D., & Vleck, C.M. (1986). Gas exchange in the incubation mounds of megapode birds. *Journal of Comparative Physiology B*, 156, 773-782.

4. Liliana D'Alba, Jones, D.N., Badawy, H.T., Eliason, C.M. & Shawkey, M. D. (2014). Antimicrobial properties of a nanostructured eggshell from a compost-nesting bird. *Journal of Experimental Biology*, *217*(7), 1116–1121. doi https://doi.org/10.1242/jeb.098343

5. Madhusoodanan, J. (2014). Nanoparticles make turkey eggs tough to crack. *Nature*, 11 April 2014. https://doi.org/10.1038/nature.2014.15039

6. Vleck, D., Vleck, C.M. & Seymour, R.S. (1984). Energetics of embryonic development in the megapode birds, mallee fowl *Leipoa ocellata* and brush turkey *Alectura lathami*. *Physiological Zoology*, 57, 444-456.

7. Jones, D.N. & Göth, A. (2008). *Mound-builders*. CSIRO Publishing. AND Booth, D.T. & Jones, D.N. (2002). Underground nesting inthe megapodes. *In* Deeming, D.C. (ed.). *Avian incubation: behaviour, environment, and evolution*. Oxford University Press. pp. 192–206.

8. Göth, A. (2007). Incubation temperatures and sex ratios in Australian Brush-turkey *Alectura lathami* mounds. *Austral Ecology*, 32, 378-385.

9. Eiby, Y. (2009). *Incubation biology of the Australian Brush-turkey (Alectura lathami)*. PhD Thesis, School of Biological Sciences, The University of Queensland. AND Eiby, Y., Booth, D. (2008). Embryonic thermal tolerance and temperature variation in mounds of the Australian Brush-turkey (Alectura lathami). *The Auk,* 125, 594–599. doi:10.1525/auk.2008.07083

10. Jones, D.N. (1988). Hatching success of the Australian Brush-turkey *Alectura lathami* in South-East-Queensland. *Emu*, 88, 260-263.

11. Göth, A. (2007). Mound and mate choice in a polyandrous megapode: females lay more and larger eggs in nesting mounds with the best incubation temperatures. *The Auk*, 124, 253-263.

12. Göth, A. & Evans, C.S. (2004). Egg size in Australian brush-turkey *Alectura lathami* hatchlings predicts motor performance and postnatal weight gain. *Canadian Journal of* Zoology, 82, 972-979.

13. Göth, A. (2007). Mound and mate choice in a polyandrous megapode: females lay more and larger eggs in nesting mounds with the best incubation temperatures. *The Auk*, 124, 253-263.

14. Dekker, R.W.R.J. & Brom, T.G. (1990). Maleo eggs and the amount of yolk in relation to different incubation strategies in Megapodes. *Australian Journal of Zoology,* 38, 19-24.

15. Göth, A., Eising, C.M., Herberstein, M.E., & Groothuis, T.G.G. (2008). Consistent variation in yolk androgens in the Australian Brush-turkey, a species without sibling competition or parental care. *General and Comparative Endocrinology,* 155(3), 742-748. DOI: 10.1016/j.ygcen.2007.11.004

Aboriginal knowledge

1. Smithers, D. (15 November 2023). Personal communication. 500voices.com.au

2. Etheridge, R. Jnr. (2011). *The Dendroglyps or 'Carved Trees' of New South Wales*. Sydney University Press.

3. Fuller, R.S., Norris, R.P., Trudgett, M. (2013). The Astronomy of the Kamilaroi People and their Neighbours. arXiv:1311.0076 .

4. FineArtAmerica. (n.d.). Brush-turkey Clifford Madsen. https://fineartamerica.com/featured/brush-turkey-clifford-madsen.html

5. Atlas of Living Australia. (n.d.). Australian Brush-turkey. https://bie.ala.org.au/species/Alectura_la thami

6. ABC Radio. (2012). *Scrub turkeys don't taste like chicken.* (Podcast). https://www.abc.net.au/local/a udio/2012/05/25/3570228.htm

7. Gray, G.R. (1861). List of species composing the family Megapodiidae, with descriptions of new species, and some account of the habits of the species. *Proceedings of the Zoological Society London,* 1861, 288–296.

8. Sorenson, E. (1919). *Friends and Foes in the Australian Bush.* Reprinted by Project Gutenberg. https://gutenberg.net.au/ebooks15/1500121h.html#ch14

9. Australian Art Network. (2021). *Surrk Au Ngarr.* https://australianartnetwork.com.au/shop/region/torres-strait-islands/surrk-au-ngarr-bush-turkey-leg/

10. Göth, A. & Booth, D. (2004). Temperature-dependent sex-ratio in a bird. *Proceedings of the Royal Society London B – Biology Letters,* 1, 31-33. doi: 10.1098/rsbl.2004.0247.

11. Göth, A. (2007). Incubation temperatures and sex ratios in Australian Brush-turkey *Alectura lathami* mounds. *Austral Ecology,* 32, 378-385.

12. New Scientist (December 2004). *Cold turkeys have fewer female chicks.* http://www.newscientist.com/channel/life/mg18424763.200. AND ABC News in Science. (2004). *Chicks take the heat, end up girls.* http://www.abc.net.au/science/news/stories/s1250633.htm

13. Eiby, Y.A., Wilmer, J.W., Booth, D.T. (2008). Temperature-dependent sex-biased embryo mortality in a bird. *Proceedings of the Royal Society B. Biological Sciences,* 275, 2703–2706. doi:10.1098/rspb.2008.0954

14. Planet Corroboree. (2020). *Wollumbin, the warrior chief & the turkey.* https://planetcorroboree.com.au/blogs/culture-country/wollumbin-the-warrior-chief-the-turkey

Other Brush-turkey studies

1. Gould, J. (1842). On the habits of the *Alectura lathami. Tasmanian Journal of Natural* Science, 1, 21-24.

2. Latham, J. (1824). *General History of Birds. Vol. 10.* Leigh & Sotheby, London.

3. Olsen P. (2010). New Holland Vulture (Australian Brush-turkey). In: *Upside down world: early European impressions of Australia's curious animals.* National Library of Australia Canberra. p. 126–131.

4. Marchant, S., Higgins, P.J. (eds). (1993). *Alectura lathami* Australian Brush-turkey. In: *Handbook of Australian, New Zealand & Antarctic Birds. Vol. 2: raptors to lapwings.* Oxford University Press, p. 341–353.

5. Hubbard, E.T. (1908). Emus and Brush-turkeys in England. *Emu,* 8, 151-152. doi:10.1071/MU908149h

6. Author unknown. (1933). Nest-Building of Brush-Turkeys. *Nature,* 131, 233. https://doi.org/10.1038/131233a0

7. Sorenson, E. (1919). Friends and Foes in the Australian Bush. Reprinted by Project Gutenberg. https://gutenberg.net.au/ebooks15/1500121h.html#ch14

8. Russel, A. (1944). *Bush Ways*. Australian Publications, Sydney.

9. Fleay, D.H. (1937). Nesting habits of the Brush Turkey. *Emu*, 36, 153-163.

10. Baltin, S. (1969). Zur Biologie and Ethologie des Talegalla-Huhnes (*Alectura lathami* Gray) unter besonderer Berücksichtigung des Verhaltens während der Brutperiode. *Zeitschrift für Tierpsychologie*, 26, 524-572. AND Jacobi, E.F. (1970). Die Zucht von Tallegallahühnern (*Alectura lathami* Gray) mit elektrischer Bruthitze. *Zoologischer Garten*, 39, 129-132.

11. Jones, D.N. (2023). *Curlews on Vulture Street*, New South Books. https://www.newsouthbooks.com.au/books/curlews-on-vulture-street-171998/

12. Jones, D.N., Dekker, R.W.R.J., & Roselaar, C.S. (1995). *The Megapodes*. Oxford University Press.

13. Jones, D.N. & Göth, A. (2008). *Mound-builders*. CSIRO Publishing.

14. https://darryljonesnature.com/contact/

How can I deter them?

1. https://www.smh.com.au/lifestyle/the-lesson-i-learnt-from-an-unwelcomed-bush-turkey-20180126-h0oktp.html

2. https://www.smh.com.au/national/nsw/fowl-play-who-s-poisoning-brush-turkeys-on-sydney-s-lower-north-shore-20230926-p5e7ot.html

3. https://www.environment.nsw.gov.au/licences-and-permits/wildlife-licences/licences-to-control-or-harm/licence-to-harm-native-animals

4. https://www.qld.gov.au/environment/plants-animals/wildlife-permits/permit-types/moving-wildlife

5. Boynton, S. (1986). Don't let the turkeys get you down. Workman Publishing Company.

How can I help?

1. Warnken, J., Hodgkison, S., Wild, C., & Jones, D.N. (2004). The localised environmental degradation of protected areas adjacent to bird feeding stations: A case study of the Australian brush-turkey Alectura lathami. *Journal of Environmental Management*, *70*(2), 109-118. https://doi.org/10.1016/j.jenvman.2003.11.002

2. https://www.sydney.edu.au/news-opinion/news/2020/11/12/track-cockatoos---bin-chickens--and-brush-turkeys-for-science.htmlAND

TABLE OF FIGURES

INDEX

www.ingramcontent.com/pod-product-compliance
Lightning Source LLC
Chambersburg PA
CBHW041258040426
42334CB00028BA/3062